# EcoMind

## Also by Frances Moore Lappé

*Aid As Obstacle (with Joseph Collins and David Kinley)*

*Betraying the National Interest*
*(with Rachel Schurman and Kevin Danaher)*

*Democracy's Edge: Choosing to Save Our Country*
*by Bringing Democracy to Life*

*Diet for a Small Planet*

*Food First: Beyond the Myth of Scarcity*
*(with Joseph Collins and Cary Fowler)*

*Getting a Grip: Clarity, Creativity and Courage*
*for the World We Really Want (2nd ed.)*

*Great Meatless Meals (with Ellen Ewald)*

*Hope's Edge: The Next Diet for a Small Planet*
*(with Anna Lappé)*

*Mozambique and Tanzania: Asking the Big Questions*
*(with Adele Beccar-Varela)*

*Nicaragua: What Difference Could a Revolution Make?*
*(primary author, Joseph Collins; with Paul Rice)*

*Now We Can Speak (with Joseph Collins)*

*Rediscovering America's Values*

*Taking Population Seriously (with Rachel Schurman)*

*The Quickening of America: Rebuilding Our Nation,*
*Remaking Our Lives (with Paul Martin Du Bois)*

*What to Do After You Turn Off the T.V.*

*World Hunger: Twelve Myths*
*(with Joseph Collins, Peter Rosset, and Luis Esparza)*

*You Have the Power: Choosing Courage in a Culture of Fear*
*(with Jeffrey Perkins)*

# Praise for *EcoMind*

"Powerful and inspiring, *EcoMind* will open your eyes and change your thinking. I want everyone to read it."          —JANE GOODALL

"This book is pivotal in the most literal sense. As I read it, I find myself turning the crucial 180 degrees from frustration and fear to a sense of constructive possibility. Frances's ability to express the most complex, existential yearnings is epic—matched only by her courage. Nothing I can say will do justice to how this book continues to affect me."

—MOLLIE KATZEN, author of the *Moosewood Cookbook*

"Lappé's effervescent enthusiasm still inspires."

—*Publishers Weekly*

"Frances Moore Lappé's exceptionally thought-provoking book is a message of hope. It shows how change is possible, once we open our eyes, look around, and see that we depend on others and on nature. This book obliges us to re-imagine our world, brick by brick, by first re-imagining ourselves."

—OLIVIER DE SCHUTTER, the UN Special
Rapporteur on the Right to Food

"I have been a social worker for almost 50 years now and I am using *EcoMind* in my Social Welfare Policy Course because it is one of the most important books that has come out during the past half-century. *EcoMind* has many important features; however, one of the most important strengths of the book is that it provides a positive way of dealing with all the negatives in the world today."

—DR. CHARLES FROST, Professor,
Middle Tennessee State University

"*EcoMind* reminds us that the most important resource for restoring a clean and healthy planet is the one sitting between our ears. Frances Moore Lappé brilliantly challenges the negative 'thought traps' of doom-and-gloom environmental messages and emerges with a positive, people-powered approach."

—MICHAEL BRUNE, Executive Director, Sierra Club

"The concepts laid out in well-organized chapters are worth revisiting for veteran activists, or discovering anew for those who have shied from the subject . . . An accessible introduction to the psychology of this 'historic challenge,' providing an enthusiastic shove toward reflection."

—*Kirkus Reviews*

"Lappé shows how by seeing the big picture we can change it. It's a clarion call in this rising age of rising despair."

—PETER BARNES, author of *Capitalism 3.0: A Guide to Reclaiming the Commons*

"Frances Moore Lappé has done it again. As she has done so insightfully with respect to food, hunger, and democracy, Lappé now turns her sights on the contemporary ecological crises. Her accessible and provocative analysis demonstrates how the ways many people think and talk about these crises—especially the dominant narratives of scarcity—obscure the inequalities of power that lie at the root of these crises and inhibit rather than inspire the kind of effective movements necessary to confront them. *EcoMind* is a profound example of how analysis breeds not paralysis but rather informed and inspired action, and is on track to do so in the 21st century just like *Diet for a Small Planet* and *Food First* did in the 20th."

—JOHN GERSHMAN, Clinical Associate Professor, Robert F. Wagner Graduate School of Public Service, New York University

"Frances Moore Lappé brings us yet another gift in *EcoMind*. She cautions us to avoid the mental traps that block our thinking. She awakens us to our immense possibilities and potentials. She invites us to release our latent energies to be the change we want to see."

—VANDANA SHIVA, Ph.D, philosopher scientist, activist and author of *Staying Alive: Women, Ecology and Development*

"[*EcoMind*] provides a fine survey that presents evidence that human beings see the world through a filter of basic belief patterns—and believe that there's not enough of anything. . . . a fine pick for a range of general-interest and science holdings alike."　　　　　—*California Bookwatch*

"Passionate and very thought-provoking. . . .[Lappé's] idea of thinking about the world not by a scarcity of eco-systems but one of plenty is, quite frankly, breathtaking in its simplicity."

—*Publishers Weekly,* Shelf Talker blog

"Brimming with useful information and analysis pertaining to developing renewable energy and eradicating waste, Lappé's lucid extrapolation of the core lesson of ecology, that everything is connected, also offers galvanizing dissections of the intensifying influence of corporations on government and the derailment of democracy. Equally compelling are her insights into how accelerating wealth inequality contributes to environmental degradation as well as poverty and why struggling people support 'policies that hurt them.' Lappé's recalibrated guide to becoming ecominded affirms our 'capacity to rethink our world' and provides many urgent reasons to do so." —*Booklist*

"The message of [*EcoMind*] . . . is still relevant." —*Santa Cruz Weekly*

"In a nutshell, *EcoMind* is like Thomas More's utopia, except that its utopia is achievable within the context of modernity, by just doing things differently, and by having a different mindset which looks at the whole integrated—and not fragmented—reality before taking a decision. Congratulations for a most readable book."
—PUSHPA M. BHARGAVA, Founder-Director, Centre for Cellular and Molecular Biology and winner of the Padma Bhushan, one of the highest civilian awards in India

"I am using *EcoMind* to teach Introduction to Environmental Studies, and finding it a terrific resource . . . making a clear argument for an approach to saving the world that is hopeful as well as ecological."
—JOHN C. BERG, Professor in the Department of Government, Director of the Graduate Program in Political Science, Suffolk University

"Frances Moore Lappé has long served as a powerful voice for food justice and a more sustainable future. Her new book *EcoMind* offers an insightful and inspirational ecology of hope, and is a must read for those concerned about the fate of the planet."
—DR. DANIEL FABER, Professor of Sociology at Northeastern University and Director of the Northeastern Environmental Justice Research Collaborative

"Well organized and filled with quotable summaries and real-world examples, this book uniquely captures how our society views itself. . . .Valuable as a general work on self-perception and the motivation to action, and essential to those feeling powerless in the struggle to reduce our environmental impact. Strongly recommended." —*Library Journal*

"[Lappé] is keenly aware of the need to weave rhetorical craft, emotional openness, and intellectual rigor into hard questions—this has been her approach since *Diet for a Small Planet* was first published in 1971 (a book that remains current 40 years later), and is the thread that connects her work in many areas, including international aid, democracy, empowerment, and of course food systems."

—RICHARD L. WALLACE, Professor and Chair of
Environmental Studies, Ursinus College

"The latest book from environmentalist Frances Moore Lappé—author of the bestselling *Diet for a Small Planet*—could not have been published at a better time . . . *EcoMind* should not be read once and stuck on a bookshelf. It should be shared with friends, discussed and challenged with others who are ready to make positive and enduring changes within their communities."

—Ms. Magazine blog

# EcoMind

Changing the Way We Think,
to Create the World We Want

Frances Moore Lappé

NATION
BOOKS
New York

*For Ida, Josie, and Rosa and your grandfather*
*Marc's courageous vision*

Books published by Nation Books are available at special discounts for
bulk purchases in the United States by corporations, institutions, and other
organizations. For more information, please contact the Special Markets
Department at the Perseus Books Group, 2300 Chestnut Street, Suite 200,
Philadelphia, PA 19103, or call (800) 810-4145, ext. 5000, or
e-mail special.markets@perseusbooks.com.

Designed by Linda Mark
Type set in Leawood by the Perseus Books Group

The Library of Congress has catalogued the hardcover as follows:
Lappé, Frances Moore.
  EcoMind : changing the way we think, to create the world we want /
Frances Moore Lappé.
    p. cm.
  Includes bibliographical references and index.
  ISBN 978-1-56858-683-0 (alk. paper)—ISBN 978-1-56858-689-2 (e-book:
alk. paper)
    1. Environmentalism—Psychological aspects.  2. Sustainable
development—Psychological aspects. 3  Environmental psychology. I. Title.
  GE195.L375 2011
  304.2'8—dc23
                                                                2011028436
ISBN 978-1-56858-743-1 (paperback)

10 9 8 7 6 5 4 3 2 1

# CONTENTS

# WHY I WROTE
# THIS BOOK

A FEW YEARS AGO I WAS ASKED TO SPEAK IN WASHINGTON, DC, AT A BIG conference on the global environmental crisis. A lot of my heroes would be presenting—nearly sixty speakers in just a couple of days.

Wow, I thought, this is bound to be just the crash course I need to make me more effective in addressing the problems I care most about. How convenient.

I did learn a huge amount in those two days. But as I walked out of the auditorium after the final speaker, something hit me. Actually, I felt that something *had* just hit me. I felt numb and heavy, very heavy. Reflecting on the experience, I noted that as the conference had worn on, the audience had wound down. I wondered what the departed ones were feeling when they left. Were they overwhelmed, stuffed so full of scary data that they felt stuck?

If others were experiencing what I was, it was not a good thing. Psychologists have found that if we believe there is no hope of overcoming a problem, many of us assume an uncaring posture to protect ourselves.[1] And if I'd learned one thing over years studying the food and environmental

Note: All URLs mentioned in this book are available on our website at www.smallplanet.org.

crises, it's that our way out of this mess is inconceivable without the active engagement of millions—well, no, billions—of us who do care. So, our earth can't afford overwhelmed, discouraged people who are too depressed to engage. _or too busy?_

Arriving home after the conference, I was deeply troubled and asked myself, Are we environmentalists actually defeating our own ends? Just when the magnitude of our environmental crises is becoming clearer by the day, are we pushing people to despair?

This question seized me.

I believe that human beings are by nature doers. Most of us love to solve problems. Without that core trait, our species could never have created our complex societies. (Forget the wheel. Forget the steam engine. Forget decoding the human genome.)

But over decades, I've also come to appreciate that central to our ability to solve a problem is how we perceive the challenge, how we *frame* it—that "seeing" determines our capacity for doing, and certainly for *effective* doing. So I asked myself, Is there a way of perceiving the environmental challenge that is at once hardheaded, evidence based, *and* invigorating—one that welcomes us to become engaged problem solvers? Might it be possible to transform something that can feel so frightening as to make us go numb into a challenge so compelling that billions of us will eagerly embrace it?

Soon I was searching for answers to that question. I began dissecting the core assumptions behind dire media messages and texts and, more broadly, those ideas relevant to the environmental challenge that float, unnamed but potent, in our culture.

Within a couple of months I'd stirred up the guts to test the water. I decided to try out what I'd been learning on participants at a "relocalization" conference in central Massachusetts. As I knew I was questioning the "no-growth," "consumerism-is-the-problem" messages dear to the hearts of many in my audience, my nervousness—the perspiring brow—was impossible to hide. But to my surprise, and huge relief, the audience responded with excitement. They peppered me with many great questions, pushing me on.

That speech became the seed of this book, which first sprouted in the fall of 2009. That's when I decided to do something I'd never before

thought of: ask my readers for help. Not only was my plate full with another book's release, but I knew my ideas were still forming. I wasn't ready
to publish a "final" version of my ruminations. I also recognized that the
essence of that first speech, and this book, was not the "facts" of the environmental crisis, in which it is the voices of the "experts" that matter. The
discussion here has to do with the way we think about the facts and
therefore what we *do* with them. So it's something to which anyone could
usefully contribute.

I put out a draft of my ideas, distributed via the website of the organization my daughter, Anna Lappé, and I run—the Small Planet Institute—and
at talks I was giving. I asked readers simply, How do I make it a better
book? I had no idea what would happen, but since then readers have contributed enough comments to fill another small book. Study groups formed
to confer together on feedback, and several professors used the draft in
their classes. In all, I was profoundly moved by people's generosity—their
willingness to give their time and effort. This is, in today's lingo, truly a
"crowd-sourced" book.

Not only did I—that is, the book—benefit enormously from the insights
of readers, but with their help I became even more convinced that what
had so deadened my spirit after that environmental conference *could* be
transformed. We don't have to keep telling ourselves a story that robs us
of the energy we need now, more than ever. We can each make the "leaps
of mind" that move us from discouragement to an empowering stance.
We can each reframe our thinking and seeing in ways that give us energy
to engage.

Get ready.

# Our Challenge—
# Developing an EcoMind

*"So where are we going? And why are we in a handbasket?"*

S EEING THIS BUMPER STICKER ON MY WAY HOME ONE EVENING, I chuckled aloud. "Wait," I thought, "that's what I'm trying to figure out." It sure seems like the question we'd all want to answer.

After all, our earth is now warmer than it's been in 650,000 years, and MIT scientists tell us that our planet's future heating will likely be twice as severe as estimated less than a decade ago.[1] So, in this century, higher water temperatures and melting ice caps could raise the sea level by nearly three feet. That's enough to flood many of the world's great coastal cities and to inundate much of Bangladesh. A rise of six feet—maybe more—is possible, and with superstorm Sandy, we already know what that feels like.[2]

But "warming" doesn't really capture what's happening. Our climate is becoming more chaotic. Think Los Angeles hitting a record 113 degrees in the fall of 2010, then a few months later Oklahoma's wind chills sinking to 31 degrees below. Or monsoon rains swelling the Indus River in 2010 to forty times its normal volume, flooding one-fifth of Pakistan's land and displacing millions.[3] Or Australia in 2006 suffering its worst drought in 1,000 years, only to face flooding over an area the size of Texas just four years later.[4]

Making climate more chaotic, each year, from Africa to Latin America, burning and logging destroy forests that cover an area the size of Greece—with climate-disrupting emissions greater than those from all transportation.[5] Partly as a result, we already may, or soon will, have wiped out enough species that the planet would need 10 million years to re-establish the extent of today's diversity.[6]

Yet, worldwide we keep on releasing more, not less, climate-disrupting carbon, with coal—by far the worst offender—growing much faster than other fossil fuels.[7]

At the same time, we're still reeling from a global financial crisis, with high rates of joblessness and worsening inequalities along with escalating food prices: In just one decade, the World Food Price Index has doubled, hurting the hungry the most.[8] Even in 2009, Andrew Simms of the New Economics Foundation in London, worrying about his country's dependence on imports, warned that "we could literally be nine meals from anarchy and we are still in denial."[9]

And here in the US, all the above can feel more daunting when the share of us who say we "worry" about climate change has dropped in recent years, now to about half, and we seem too bitterly divided as a culture to act.[10]

Are you scared? I know I am.

But I realize that's not the real question. The real question is whether we each can move ahead creatively *with* our fear because we believe that, in this pivotal moment, we have it in us to make a planetwide turn toward life.

I believe we do.

But don't get me wrong—I am not an optimist. I am a staunch, hardcore, dyed-in-the-wool possibilist. I believe it is possible that we can turn today's breakdown into a planetary breakthrough—on one condition: We can do it if we can break free of a set of dominant but misleading ideas that are taking us down.

Ideas?

Yes. The poetic observation often attributed to French writer Anaïs Nin that "we don't see things as they are, we see things as we are" is precisely what scientists now confirm experimentally: For human beings there is no unfiltered reality. We are creatures of the mind who interpret experience through a largely unconscious mental map made up of the big ideas orienting our lives. Philosopher Erich Fromm called it our "frame of orientation," through which we see what we expect to see. So, while we often hear that "seeing is believing," actually believing is seeing.[11]

Revealing this deeply human trait is a silly but telling experiment in which psychologists instruct subjects to count basketball passes by players wearing white. In the middle of the game, a person in a gorilla costume appears and pounds her chest directly in the subjects' line of vision; yet, a good half of them don't register this unexpected antic at all. They're focused on counting basketball passes![12]

This trait—seeing only what we expect to see—even shapes how we perceive our own nature and our place in the universe and, therefore, what we imagine to be possible. I first grasped the huge import of this trait when, as a college senior, I was assigned Thomas Kuhn's classic work *The Structure of Scientific Revolutions*. In it, Kuhn shows how difficult it is for humans to shed a reigning mental map. Even bright people clung to an earth-as-the-center-of-the-universe worldview for 150 years after Copernicus showed us that, no, the earth is not at the center, we revolve around the sun.

> To see what is in front of one's nose needs a constant struggle.
>
> —George Orwell, "In Front of Your Nose," 1946

Once we see through a certain lens, it's hard to perceive things differently, be they the most mundane matters or the most momentous. Yet, the hard fact of human existence is that if our mental frame is flawed, we'll fail no matter how hard and sincerely we struggle.

The central problem this book addresses is that, sadly, much of humanity today is stuck in precisely this "hard fact"—trapped in a mental map that defeats us because it is mal-aligned both with human nature and with the wider laws of nature. So, the question is, Can we remake our mental map? And do it much faster than those early astronomers?

## CAN WE REMAKE OUR MENTAL MAP?

Before exploring this central question, let me share four observations that bolster my cockeyed possibilism.

## One: We're living an aberration

It's not always been this way. Much of the systemic destruction we're now experiencing is a great and brief aberration.

If all human history were squeezed into one week, and the clock started on Monday, our industrialized world—spanning only about *seven* generations—would emerge at three seconds before midnight on Sunday.[13] In the one hundred years of the twentieth century, humans used ten times more energy than we did in the previous 1,000 years.[14] In fact, 60 percent of the increase in atmospheric carbon dioxide levels now threatening our planet has occurred just since I was a high school freshman in 1959.[15]

So what we are experiencing may be horrific, but it is not the norm of human experience. It is not "conventional" or "business as usual." Let's banish the terms. It is rather a huge and failing experiment, a sudden, radical detour.

## Two: We already know how

Solutions to our crises—from global climate chaos to global hunger—are largely known. Consider this quick scan of four of our biggest challenges.

*Starting with the energy-and-climate crisis . . .*

While planet-heating coal now supplies about half of US electricity, renewable energy—wind, solar, geothermal, bioenergy, and hydropower—has the "technical potential" to provide more than sixteen times the electricity the United States needs now, concludes the Union of Concerned Scientists' "blueprint" for getting to a green economy. In fact, the study tells us, any *one* of three green sources—wind, solar, or geothermal—could meet current electricity needs.[16]

So, by tapping even a portion of this potential, we could replace coal.

Just two quite doable steps—raising fuel-economy standards and improving home and industry energy efficiency—could, over a twenty-five-year period, save the United States more than 3 billion barrels of oil a year, the same report notes. That's nearly half what we consumed in 2009.[17]

In a different 2004 study partially funded by the Pentagon, physicist Amory Lovins explains how it's possible to wean the US economy off oil in a few decades, mainly through greater efficiency and a shift to green energy sources. He shows that by investing an average of $18 billion a year over the course of a decade—that's less than 14 percent of what we were recently spending on average in Iraq and Afghanistan each year—we'd realize a net savings of $70 billion a year by 2025.[18] Plug in 2011 oil prices and our projected savings would soar.

These projections also show that along the way, we'd revitalize industries, create more jobs, and make the US more secure than we would if we'd stayed the fossil fuel course to its bitter end.

One reason, Lovins persistently reminds us, is that saving a barrel of oil is a whole lot cheaper than buying one. At this writing a barrel of oil costs about $100, but saving a barrel costs only $18.[19]

Getting off oil in just a few decades? Have Americans—or anyone—ever moved this quickly?

The answer is yes.

Even if Americans began saving energy at only *two-thirds* the pace we achieved when reacting to the oil price shock from 1977 to 1985, we could be off oil in thirty to forty years, Lovins estimates.[20] Other countries are already speeding along this path. Consider Sweden. By 2009 it was already getting more of its energy from biomass—plant material—than from oil.[21]

Costa Rica—which discovered oil but in 2002 placed a moratorium on its exploitation—now gets 95 percent of its electricity from renewable sources—hydroelectric, wind, biomass, and geothermal. But Costa Rica isn't satisfied. It is rushing to become the world's first carbon-neutral country in time for its bicentennial in 2021, says Minister of Environment and Energy Roberto Dobles. Four other countries are close on its heels: Monaco, Norway, New Zealand, and Iceland.[22]

When mulling over what's possible, I also feel fortified by noting that countries now emitting vastly less carbon per person than the US are at the same time great places to live. Shouldn't the fact that Germany releases half as much carbon dioxide per person as we do strengthen our confidence that we can get there and beyond?[23]

Plus, note that even the experts have way underestimated what's possible: A recent survey of nearly fifty forecasts in Europe and around the world discovered that "nearly all of them had underestimated the future increases" in renewable-energy generation we would in fact achieve.[24] A few years ago, for example, the International Energy Agency set an ambitious goal for world wind-energy capacity for 2020—a goal we surpassed more than a decade early.[25] One reason is that, by 2009, the US (led by Texas!), China, and Germany had together installed more wind power than the rest of the world combined.[26]

*And deforestation . . . do we know how to stop it?*

Felling and burning the earth's forests massively accelerates the pace of climate disruption. But compared to the 1990s, the next decade saw the earth's net loss of forest—though still horrific at 13 million acres annually—drop by more than a third. Even Indonesia, with one of the worst rates of deforestation during the 1990s, reduced its rate of loss.[27] In India, the government shifted from top-down forest management, enabling forest management by tens of thousands of village forest-protection groups.[28] Its forests improved and expanded, and over the last decade, India ranked among the world's top ten countries by yearly net increase in forest area. In 2005 it also led the world in area reforested.[29]

*Or take food and farming.*

We know how to get that right, too, even though we've gotten it *really* wrong for sixty years: Extractive, destructive agriculture has created more than four hundred oceanic dead zones worldwide, where farm-chemical runoff is devastating aquatic life.[30] (One dead zone in the Gulf of Mexico is often as large as the state of New Jersey.)[31] And today, the global food and agriculture system—largely due to its increasing chemical intensity, the growth of the grain-fed livestock industry, and forest clearing for farming and grazing—generates roughly one-third of the climate impact of greenhouse gas emissions.[32]

At the same time, evidence mounts that we don't need to wreak havoc to feed ourselves well. Think jubilant farmers in Mali, using nonchemical practices, who in 2009 won a prize for rice yields more than double the world average.[33] And a number of studies now confirm the exciting promise of these farmers' ecological approach.

An extensive 2007 University of Michigan study, for example, estimates that moving globally to organic, ecologically attuned farming practices could increase food output significantly.[34] The shift is already saving and transforming the lives of millions, even in ostensibly resource-poor areas: In twenty African countries, more than 10 million farmers have on average doubled their yields by adopting agroecological approaches such as composting, mulching, and careful intermixing of crops, according to recent research sponsored by the UK government's Office for Science. Their farms cover an area more than half the size of the UK.[35]

Other evidence of possibility?

Compared to industrial farming, organic methods generate one-half to as little as one-third as many greenhouse gas emissions.[36] In a decade organic agricultural land has tripled, and by moving worldwide to organic practices in two decades agriculture could be carbon neutral—releasing no more than it's absorbing—says the UN Environmental Program.[37]

*Finally, we know how to end hunger and poverty too.*

Here at home, we were well on our way from the late 1940s to the early 1970s. Over these few decades, the poorest fifth of Americans saw their real family income jump 116 percent, the biggest leap of any income group.[38] Achieving this striking progress were pretty straightforward poverty-fighting strategies: taxation based on ability to pay, a labor movement covering a third of private workers, high rates of employment, public support for veterans' education, a minimum wage packing 25 percent greater buying power than it does today, and more. They all added up. In just over one decade—the 1960s through the early 1970s—we cut our poverty rate in half.[39]

The US has since gone backward fast—with almost 11 million more people sinking into poverty over the last decade and child poverty rising to 20 percent. Others have not. Seventeen years ago almost one-third of children in the UK lived in poverty. But the Brits raised child welfare benefits and kept their minimum wage much higher than ours. These and other efforts slashed the child poverty rate by more than half, to 12 percent.[40]

Even more dramatically, in the Global South, Brazil in just six years—from 2001 to 2007—cut poverty by 25 to 40 percent, depending on one's measuring tool. Similiarly, Vietnam cut its poverty rate from 58 percent to 16 percent in less than two decades using public investment in jobs, education, housing and more.[41] In part Brazil's success reflects Bolsa Família, a cash bonus introduced in 2003 that goes to poor families that keep their kids in school and make sure they have medical care—a huge boon to families whose survival might otherwise depend on their children's labor.[42]

Whether in a poverty-plagued Latin America or in the big, rich USA, commonsense strategies have worked to make advances against hunger and poverty.

So yes, we are in big trouble, *but it's not for want of answers*. This is the second reason I'm a possibilist. Solutions are known and are within our reach.

## Three: It's not all locked up

Surely one reason it's easy to feel defeated is that we're not hearing about striking advances like these. Yet another is a common perception that the global power balance is so skewed that, in effect, "it's all locked up." We can feel shut out by an intimidating global corporate stranglehold whose grip, not our actions, feel all-determining. With corporate logos slapped on everything from tacky T-shirts to treasured public places—think Dunkin' Donuts Civic Center or Cisco Field—it's easy to feel that our planet is now strictly in global, corporate hands.

Here, too, our sights widen to possibility if we consider that giant corporations are not the only players in our economies.

While it's true, for example, that a handful of corporations—Cargill, ADM, and Bunge—do dominate the global grain trade, it is also true that more than 85 percent of the world's food is consumed in the country where it's grown, according to UN agricultural data. Often it is sold within the same region—much of it outside the formal market system. And it turns out that most of world agricultural production isn't the work of agribusiness, but of half a billion small farms, says the UN Environmental Program. We can also thank pastoralists, hunters and gatherers, and let's not forget the 800 million urban and near-urban farmers and gardeners.[43] In Japan's metropolitan areas, for example, there are 2.5 million acres of farmland plots annually producing food valued at $28 billion.[44]

So, to conceive of small producers as "marginal" is quite a stretch. They play a central role in food production.

And jobs?

In Latin America, the street vendors, urban food growers, craftspeople, and service providers of the informal economy created 85 percent of the jobs in the 1990s and roughly 50 percent in the last decade.[45] And in India, despite the media's focus on high tech, nearly 90 percent of Indians work in this informal economy.[46] By contrast, information technology and outsourcing employ only a few million people in a population that exceeds 1 billion.[47]

To see the world economy from a more realistic and empowering perspective, also note it's likely that more people are members of cooperatives—one person, one vote—than own shares in publicly traded companies, based on one dollar, one vote.[48] Cooperatives also provide one-fifth more jobs worldwide than do multinational corporations.[49]

My point is neither to glorify the Global South's often harsh, even horrific working conditions nor to imply that small producers aren't affected by global marketing and processing corporations. It is simply to remind us that our economies are not all sewn up by centralized corporate structures. Even in the US, businesses with fewer than five hundred employees still produce roughly half of the gross domestic product that is private and not from farming.[50]

## Four: And a lot of people care

Finally, our problem is not a disinterested citizenry. Hardly.

While, as noted above, the share of Americans who say they "worry" about climate change has fallen, survey after survey shows widespread concern and desire for action. Even as the economic crisis hit in 2009, four in ten Americans still ranked the environment as a top priority.[51] Just a few years ago, nearly 80 percent of us said we were "ready to make significant changes to the way [we] live to reduce climate impact." And about 70 percent of people polled in twenty-one countries agreed.[52]

In 2010, a Stanford poll found in the US that 86 percent of respondents wanted the federal government to act to limit air pollution from businesses, and 76 percent wanted legislation to limit the greenhouse gases that businesses can emit.[53] And even though almost half of us believe (falsely) that there's a trade-off between economic well-being and our environment, two-thirds of likely US voters in early 2011 agreed that renewable energy is a better long-term investment for our country than fossil fuel.[54] Finally, two-thirds of us agreed, in response to a 2009 survey by the Glaser Foundation, that "America must play a leading role in addressing climate change . . . complying with international agreements on global warming."[55]

This is the fourth reason I'm a possibilist. Despite hand-wringing about our political divide, there's a lot of evidence, documented throughout this book, that Americans yearn to be part of the solution.

## SO WHY ARE WE MOVING BACKWARD?

If answers seem to be right in front of our noses, and our global economic reality isn't as locked down as it can seem, and many people do care, *what's our problem*?

It's that too many of us feel powerless.

This is what we *really* have to worry about—for what good are proven answers if we don't have the power to manifest them? If we can't see how our individual acts can possibly count, given the enormous clout of those invested in the current course?

Almost nine in ten of us feel "big companies have too much power and influence in Washington," and lobbyists and political action committees don't fare much better in the public mind. When it comes to media, two-thirds of Americans share a basic distrust.[56] We feel overpowered, dismissed, shut out of our home—democracy.

If you're with me to this point, the next question is pretty obvious: How do we become powerful? How do we discover and build our power to create democratic decision making that responds to *us*?

## THE POWER OF IDEAS

I approach the answer this way: To get a grip on what's robbing us of power, I ask myself, Who or what could be powerful enough to keep us creating, as societies, a world that as individuals we abhor—a world violating our deepest sensibilities and common sense? Over the decades, I've become convinced that the answer is not "those bad guys," whether they are officials in Washington with whom we disagree or those threatening us from caves in Afghanistan.

As you now know, I see only one force that potent: the emotional power of our own ideas to trap us or to free us. This human quality of seeing the world through a particular lens is perfectly fine, so long as the ideas shaping our reality serve life. But what if they don't? What if today, as our planet faces unprecedented threats, several dominant ideas—like the once tenaciously held notion that the sun circles the earth—aren't serving us well at all?

For me, these dangerous ideas, together making up a coherent world-view, begin here:

At their core is the premise of lack, the notion that there just isn't enough—of *anything*. Not enough food or fuel, jobs or jungles, parking spots or pandas, laughter or love. In fact, modern economics, now a dominant world religion, defines itself as the science of allocating scarce goods. And, unfortunately, even many environmentalists reinforce this view. In a recent call to action by environmentalists I admire, I read that all the stuff we use is made from something "scarce" that came from the earth and is produced by "scarce energy" from fossil fuel.

But perhaps even more debilitating than the notion that there just aren't enough goods for our well-being is a parallel assumption: There isn't enough goodness either. Our culture seems to whittle the human essence down to a caricature: We are selfish, materialistic, and competitive. At least, that's all we can truly count on, and the way we've always been. Thus, even in accepting the Nobel Peace Prize, President Barack Obama informed the world—while contradicting archeological evidence—that "war, in one form or another, appeared with the first man."[57]

So, the worldview we absorb everyday is driven by a *fear of being without*—without either the resources or human qualities we need to make this historic turnaround. Within this Western, mechanical worldview that we absorb unconsciously, we are each separate from one another, and reality consists of quantities of distinct, limited, and fixed things. I think of it as the three *S*'s: separateness, scarcity, and stasis. That's our world.

## THE GIPPER AND GEKKO

And what does this worldview look like when, in pure form, it drives a society?

In our country, in just one generation it has emerged in a philosophy that denigrates the public sphere—as in 1981 when Ronald Reagan, "the Gipper," declared in his first inaugural address that "government is the problem"—while it celebrates individual self-seeking. Recall Gordon

Gekko's infamous "greed is good" line in 1987's *Wall Street*? It captured for many Americans the spirit of the era, just as soaring sales of Ayn Rand's me-first novels do today.

A result is an accelerating concentration of wealth, becoming so extreme that by 2005 Citigroup had named our economy a "plutonomy" because 1 percent of households control more wealth than the bottom 90 percent.[58]

The worsening stress and deprivation brought down on the majority of us only lend further credence to the worldview's core tenet: *lack*.

This philosophy—fed today both by dominant political voices and by constant commercial messages—encourages us to see ourselves in endless competitive struggle, without the innate capacities to come together in common problem solving: in other words, lacking what democracy itself requires. While reversing our downward trajectory demands effective, responsive government more than ever, we've been absorbing the notion that government *itself*—not the forces making it less and less accountable to us—is our problem. In sum, this worldview turns us not only against each other but against an essential tool we have in common to meet our common needs.

Once we distrust government, it then makes perfect sense to privatize everything we can—from schools to prisons to many aspects of war.

It wasn't always this way. Growing up, I learned in public high school in Texas that democracy entails the coming together of differing perspectives to deliberate over what is best for all and then compromising until a path is chosen. In a 1964 poll, when asked whether they trusted the federal government to do the right thing all or most of the time, three out of four Americans responded positively.[59]

Soon, however, a take-no-prisoners politics took hold that is the logical extension of a worldview of endless competitive struggle.

In it, the democratic process is without intrinsic value. It is a means to further one's pre-set ends—discarded when it gets in the way: as in early 2011, when Michigan passed legislation permitting the governor to declare a municipality in financial crisis and to appoint a manager to "act for and in the place of the governing body."[60] One approving lawmaker called it "financial marshal law."[61]

The goal of politicians, in this view, is not to win a public debate or "make a deal" to achieve a legislative solution; it is to destroy the other side. "Politics is war conducted by other means. In political warfare you do not fight just to win an argument, but to destroy the enemy's fighting ability. . . . In political wars, the aggressor usually prevails," writes David Horowitz in "The Art of Political War," a pamphlet first distributed by Republican congressman Tom DeLay to his colleagues in 2000, later turned into a book, and updated most recently for Tea Party activists.[62]

And, sure enough, politics has become more and more warlike. In this frame, blaming the other becomes standard public discourse; compromise is treason.

Just as predictably, the public's view of government reflects its ongoing denigration. By 2008, to the polling question above concerning one's view of government, less than a third of us expressed trust, a drop of about 60 percent over four and a half decades.[63] And by 2010, a CNN/Opinion Research survey found that 56 percent believed the federal government "poses an immediate threat to the rights and freedoms of ordinary citizens." Now, that's harsh! And among Independents and Republicans, about two-thirds held this damning view.[64]

## THE TRAP

These extreme attitudes appear to some as only the latest partisan political trend. But central to the thesis of *EcoMind* is that they actually reflect a much deeper set of assumptions that extend across political boundaries and affect virtually all of us. They help to explain why many accept or even endorse policies that hurt them and benefit only those at the very top of the economic ladder, such as massive cutbacks in services and the refusal to tackle the environmental crisis. My hypothesis is that many of us fall in line because the "it's *my* money" and "the individual is king" messages "click" neatly into a preformed emotional mind-set, one grounded in an assumption of lack and separateness.

It's tempting for environmentalists, including me, to imagine that we're untouched by this dominant frame of lack of "goods and goodness" with its presumed endless competitive struggle.

But we're not. In *EcoMind* I explore seven widely held environmental messages and related ideas—some of them largely unspoken assumptions—that now shape our culture's responses to the global environmental and poverty crises. In each case, I challenge their limiting premises because I believe they are still trapped in the dominant frame of lack and separateness, and I offer a reframing that I believe can help free us to find our power to create the world we really want.

## THE SEVEN THOUGHT TRAPS

**One:** Endless growth is destroying our beautiful planet, so we must shift to no-growth economies.

**Two:** Because consumers always want more stuff, market demand and a growing population drive endless exploitation of the earth.

**Three:** We've had it too good! We must "power down" and learn to live within the earth's limits.

**Four:** Humans are greedy, selfish, competitive materialists. We have to overcome these aspects of ourselves if we hope to survive.

**Five:** Because humans—especially Americans—naturally hate rules and love freedom, we have to find the best ways to coerce people to do the right thing to save our planet.

**Six:** Now thoroughly urbanized and technology-addicted, we've become so disconnected from nature that it's pretty hopeless to think most people could ever become real environmentalists.

**Seven:** It's too late! Human beings have so far overshot what nature can handle that we're beyond the point of no return. Democracy has failed—it's taking way too long to face the crisis. And because big corporations hold so much power, real democracy, answering to us and able to take decisive action, is a pipe dream.

These seven "thought traps" are offered not as a definitive list but to encourage all of us to examine, and to reshape if appropriate, the stories we tell ourselves and others. Perhaps you've heard some or all of them stated explicitly or implicitly. You may agree with them, at least in part. If so, I invite you to suspend judgment for just a moment and to consider that even seemingly commonsense ideas can be dangerous—*if* they come across in ways that trigger self-defeating emotions, if they evoke fear and despair.

Or ignite guilt.

"Like all outlaws, we're now being punished for our transgressions—and climate change is just the scariest of the retributions that may be visited upon us," writes Jonathon Porritt, called by the UK's *Guardian* "the most influential green thinker of his generation."[65] In a similar morality frame, many see the cause of the environmental crisis as the "insatiable consumer" or our "age of irresponsibility." Unfortunately, such metaphors not only make us feel blamed but fix attention on character failings. They don't help us to identify patterns of causation and the rules that create those patterns.

Dominant metaphors of much of contemporary environmentalism—like "power down" and "we've hit the limits"—coupled with our culture's more subterranean assumptions about our separation from nature—fail to crack the worldview of "lack of goods and goodness." They can feed instead the fear and division that sustain this disempowering worldview.

Moreover, they fail to offer emotionally compelling alternative ways of seeing challenges and their rich, positive possibilities. This is a huge shortcoming, since we humans are creatures of meaning: We don't jump into a meaning void: We must see a new path in order to leave the old.

*An ecological lens.*

Fortunately, there is another way of seeing now opening to us and, through it, a new pathway. We can see the world and our place in it through the lens of ecology. Ecology is, after all, simply the relationships among organisms and their environment.

With this lens we leave behind any fixation on quantities of limited things. We see that ours is not a finished, fixed world of distinct entities but an evolving and relational world. Through an ecological worldview, we realize that everything, including ourselves, is co-created moment to moment in relation to all else. In the words of visionary German physicist Hans-Peter Dürr, "There are no parts, only participants."[66]

At its deepest, this insight lies at the heart of ancient wisdom traditions as well as the newest thinking in physics, biology, and neuroscience.

Despite literature in the field of experimental neuroscience still "dominated by 'top down–bottom up thinking,'" our brains are "feeding back to and directly linking regions that were not known to communicate with one another," report University of Southern California professor Larry Swanson and colleague Richard Thompson. They are discovering, instead of top-down control, something more like "vast networks such as the internet," where there is neither top nor bottom.[67]

Echoing their insights is Oxford physiologist Denis Noble in *The Music of Life*. In biological systems, he writes, "there are no privileged components telling the rest what to do. There is rather a form of democracy [involving] every element at all levels." The shape of life, Noble explains, emerges through the interactions of all of the components of the system with each other.[68]

An eco-mind thinks . . .
• less about quantities and more about qualities;
• less about fixed things and more about the ever-changing relationships that form them;
• less about limits and more about alignment;
• less about what and more about why;
• less about loss and more about possibility.

Separateness is therefore the illusion; notions of "fixed" or "finished" are also fanciful. Mutually created and ever changing—that is reality.

In the dominant coherent, yet self-defeating, way of seeing, the "environment" is something outside of ourselves that needs help, really fast. From this standpoint, one perceives oneself as joining and enlisting others in an environmental movement to rescue the planet.

But as we rethink the premises underlying this worldview, we move to a different place altogether—a place where we experience ourselves and our species embedded in nature. We discover, for example, that not only do we exist *in* a habitat, we *are* a habitat. In our mouths alone live more than seven hundred species of bacteria, pioneering biologist E. O. Wilson informs us. And thankfully so, as they help fend off pathogens. In fact, Wilson reports, "most of the cells in our bodies are not human but bacterial."[69]

With an eco-mind, we move from "fixing something" outside ourselves to re-aligning our relationships within our ecological home.

## A LIBERATING JOLT

Of course, I'll understand if you have big doubts about whether we humans are capable of remaking our mental frames, especially since they often lie beneath our conscious awareness. So, as a reminder of our capacity to shift perspective and its consequences, I invite you right now to try a little experiment.

Lift one hand in front of your face, palm toward you, and let your fingers part slightly to allow a bit of space between them. Focus close in, just on your palm and fingers. When I do this, I see mostly my hand, and that's it. Now, ask me to observe the room without moving my hand, head, or eyes. Suddenly, I realize I can. Even with my hand in front of my face, I can see the entire room by merely shifting my focus, *if I choose*.

Similarly, we as a species may be able to shift our focus and choose a new context for viewing our world. We can see the environmental catastrophe within a vastly bigger "room"—one that connects us with all around us.

But is such dramatic change possible?

I believe we're capable of gigantic shifts of perception, including some very sudden ones. Even as I write in early 2011, Westerners' long-held perception of Middle Eastern autocrats being in firm control shattered in a matter of weeks. But what really allows us to believe in the possibility of remaking core assumptions is, of course, our own direct experience.

Sometimes it takes a huge jolt. When I was twenty-six, newspaper headlines and world hunger experts everywhere were showering us with the scary news: Human numbers were hitting the limits of the earth's capacity to feed us. Massive famine was around the corner.

Were they right? I had to know. So I began contrasting these pronouncements with the data I was digging up. And soon I sat in shock: What? Scarcity *isn't* the cause of hunger? It seemed impossible to believe, but yes, food was then, and still is, abundant. Redrawing that piece of my mental map led to new questions and more shifts of perception.

Today, I believe the majority of us are experiencing psychic dislocation, or what psychologists call cognitive dissonance, that unsettling feeling that one's world just doesn't fit together anymore. Perhaps never in human history have such waves of shock, threat, and hope—from global climate disruption to financial collapse to democratic revolutions—arrived simultaneously for so many. So my hunch is that if there were ever a moment in which big, societywide change might be possible, this is it.

At moments like these, some long-standing assumptions suddenly seem inadequate, even useless. Imagine poor Alan Greenspan, once the revered head of the US Federal Reserve, whose seeming ability to foresee changes in the financial markets had earned him the nickname "Oracle." In 2008, confronted with one of the greatest financial free falls in American history, he had to acknowledge a "flaw" in his view of how the world works.[70] What cracked, said the *New York Times*, was Greenspan's "resolute faith" that those participating in financial markets would act responsibly.[71]

A moment of dissonance can be terrifying. But it can also be a great gift—a liberating whack. As long-held blinders fall away, we can see what in "normal" times was hidden. We can choose to freeze in fear and retreat. Or we can see ourselves and the world with fresh eyes. As we make big "leaps of thought," we can move from disempowerment and despair into an upward spiral of empowerment and honest hope. With new clarity, suddenly we have real choice—maybe for the first time.

So, in this book, I probe and challenge the seven thought traps above. My hope is that this exploration can help us all to realize the most stunning implication of an ecological way of seeing: endless possibility. Now, to feel the freedom of an eco-mind in our bones, let's probe those thought traps holding us back, take some big leaps—and then explore the ground on which we land.

thought trap 1:

# No-Growth
# Is the Answer!

*Endless growth is destroying our beautiful planet,
so we must shift to no-growth economies.*

E VER SINCE THE PHENOMENAL BUZZ SURROUNDING THE PUBLICATION
of the book *Limits to Growth* written by a team of young MIT scientists almost four decades ago, this message has seemed like a no-brainer to many. Today, the dean of no-growth is indisputably former World Bank economist Professor Herman Daly—a true pioneer in green economics.

To create sustainable societies, Daly declares, we must leave behind the "growth economy" in which success is defined as ever-increasing production.[1] Striking the same note, Worldwatch Institute's Erik Assadourian warns that a no-growth economy is "essential" if "the wealthy countries . . . are to rein in carbon emissions."[2] The economies these luminaries envision, no longer disrupting nature's cycles, are where we must quickly head.

*But stopping "growth"?* Hmm.

Growth sounds pretty good to my ears, especially when I consider the alternatives: shrink, shrivel, decline, decrease, die. All these sound, well, downright unappealing. And for the majority of the world's people, those struggling without paid work or fearing layoffs, I can see why the approach could feel threatening—signaling to me that "no-growth" might not be environmentalists' most stirring rallying cry.

So, while fervently embracing the goal of ecologically benign economics, I can't quite visualize excited crowds waving their placards in the air— "No-growth NOW!"

The problem with this way of framing our reality cuts much deeper than whether it's sexy. Its first big downside is that the frame leaves unchallenged the prevailing assumption that what defines today's economy is in fact "growth"—ever-expanding abundance. It lets stand the notion that our economy has for the most part brought us great stuff; it's just too bad we can't keep going on this happy path.

This framing is a huge obstacle. It blinds us to the reality that what we've been doing actually generates much more waste and scarcity than abundance—for many now and for many more in the future.

This realization was the *aha* moment I mentioned earlier that set me on fire at age twenty-six. Squirreled away in the University of California, Berkeley, "ag" library, I was trying to piece together an understanding of why hunger still exists in our world. And in the process, I soon discovered that our "efficient, modern, productive" US food system was not creating the plenty I'd imagined. In fact, I learned, it funnels sixteen pounds of grain and soy into cattle production to get back *one* single pound of beef.[3] At first, I assumed that such a wasteful ratio had to be an exception, but gradually I came to realize that gross inefficiency is the rule. Here is what we're really producing:

> *Resource waste:* Ten tons of "active mass raw materials" such as coal and wood are extracted for each person in the US each year, reported a widely used 1989 study by economists Robert Ayres and A. V. Kneese. Yet, only *6 percent* ends up in "durable products" we use. The rest becomes waste as fast as it is extracted.[4] Plus, compared to fifty years ago in the US, we generate almost two-thirds more municipal solid waste per person—now over four pounds each day.[5]

*Energy waste:* Fifty-five percent of all energy in the US economy is wasted, reports Lawrence Livermore National Laboratory.[6] Other experts say it's even worse—with 87 percent wasted.[7] These findings are less surprising if one considers that about two-thirds of energy entering most of the world's power plants—as coal or oil, for example—is released as waste heat.[8] Some waste of energy is unavoidable, true; but we lag far behind efficiencies achieved in many other industrial economies.[9]

*Water waste:* To produce just one pound of beef in the US uses as much as 12,000 gallons of water.[10] So, by one estimate, it takes one hundred times more water to produce a unit of beef protein than the same amount of protein from grain.[11] And much of that water is mined for irrigation from America's largest aquifer (underground, water-bearing earth), the Ogallala, which lies beneath our farm belt. If we keep using the water at this rate, portions of the Ogallala could be empty in just twenty-five years.[12]

*Ocean waste:* In recent decades we've fully exploited, or overexploited, three-fourths of the world's fish stocks.[13] But for what? One-third of the catch gets turned into feed for animals, wasting much of its potential as food for people.[14]

*Food waste:* Nearly half of the food ready for harvest in the US never makes it into our bodies. Fresh fruit, for example, can deteriorate in our long supply chains, but even 14 percent of what Americans purchase is thrown out.[15] So think of it like this: Every day, besides feeding us, our food system wastes enough to meet the caloric needs of a second country about two-thirds our size; and it's getting worse—with wasted calories increasing by half since the mid-1970s.[16] To boot, most of our wasted food ends up in landfills, generating the powerful climate-disrupting gas methane.

Plus, we shrink our food supply in another way. Worldwide, about a third of grain and over 90 percent of soy meal now go to livestock, returning to humans only a fraction of the nutrients fed.[17]

Clearly, staggering waste and loss are the rule, not the exception. Yet, because we can't see most of this built-in waste and destruction, some environmentalists critical of the current order continue to describe our economy as being "designed expressly to create wealth"—which sounds wonderful.[18] And in their textbook *Ecological Economics*, Professors Herman Daly and Joshua Farley describe our earth as a "ship" whose cargo hold has been overloaded with our "gross material production" or is "nearing capacity."[19] The primary problem of our growth economy, Daly suggests, is that it may well already be generating more "physical wealth" than "the biosphere can sustain." Ours is a "full-world economy," he says.[20]

Because I share these trailblazing economists' goals, I worry that such metaphors can't work: to most people, "full" isn't something to be upset about; it sounds really good—full heart, full life, full tank! More worrisome, such quantitative images don't help develop our eco-minds so that we can see patterns of destruction.

One might more aptly argue that our planet was once "full"—full of a vast complexity of life forms, a fullness that our economies have been emptying faster and faster. Every day we lose over 100 species.[21] And rather than *creating* wealth, more fundamentally our economy is designed to *concentrate* wealth, a form of concentration involving the vast destruction of real wealth, the health of the natural world, including human life. Today, 60 percent of ecosystem functions that sustain life worldwide are "being degraded" or used in ways that can't last.[22]

Since what we've been calling "growth" is largely waste and destruction, let's call it what it is: a system that in fact stymies growth and even quickens diminution and death—of genetic and social diversity, health, relationships, beauty, happiness, art forms, languages, and ancient knowledge.

So the first big problem with the "growth-is-our-problem" frame is that it blinds us to the massive waste-making machine our economy has become.

## BUT WHY THE BUILT-IN WASTE AND DESTRUCTION?

To cure that blindness we have to get our heads around our economic system—this religion we've been spreading around much of the world—and understand why it's become the waste-generating aberration that it is.

In the blame-growth view, the answer to the question "Why?" seems simple: Our problem is the quantitative overtaxing of resources—just too much. Hence, the frame's second big downside: It inhibits us from digging, from following our curiosity.

Here's where my curiosity leads me:

Evidence suggests a basic design flaw in our peculiar version of a market. Markets have served humankind for millennia, but we've turned this useful tool into a formula for disaster—a market that ends up producing waste and destruction because it is largely driven by one-rule: Pursue what brings the most immediate and highest return to existing wealth holders.

With single-minded fixation on this end, our economy creates scarcity from plenty in four ways:

First, one-rule economics violates nature's laws, disrupting its regenerative power. Focused only on financial return to a minority, our market isn't designed to respond to other signals—nature's signals that could avert, for example, the steady loss of soil fertility, the ongoing depletion of groundwater, the 70,000 annual deaths from polluted air in the US alone, the 20,000 deaths each year worldwide from pesticide poisoning, or the multifarious consequences of climate-altering greenhouse gases.[23]

No. Our market can't even register these signals. Economists call them "externalities" because they're external to the financial balance sheet of the corporation producing them. But a corporation's "externality" is our reality.

Coal, for example, emitting 50 percent more carbon per unit of energy than oil, remains a highly lucrative industry for companies like Massey and Peabody. But that's only because this industry's "externalities"—its real costs in public health impacts and in environmental damage—are paid not by coal companies but by the rest of us. And they are vast: about a third of a *trillion* dollars each year, according to a 2011 study led by Paul Epstein of Harvard Medical School.[24]

More broadly, the externalized impact of just 3,000 of the world's biggest corporations, report two UN-affiliated bodies, totaled over $2 trillion in damage to the global environment in 2008—that's equal to most of the 2009 US federal revenue.[25]

Second, as we should have learned as kids playing Monopoly (also governed by simple rules aggregating wealth till the winner takes all), our market concentrates financial returns so tightly that most of the earth's people experience scarcity, *no matter how much we produce.*

Worldwide half of all people survive on less than $2.50 a day, while the richest four hundred Americans now control more wealth than the poorest half of all the world's adults.[26] One family—the Waltons of Wal-mart—has come to control roughly as much as the bottom 40 percent of Americans put together.[27] Yet the bottom 90 percent of Americans now make less in real dollars than in 1973—down on average $2,000.[28]

The third pitfall of a one-rule economy is that it ends up depriving us of the open, transparent, and fair public conversation that is the heartbeat of democracy. As wealth concentrates and the idea of a public good loses favor, our communications media over the last thirty years have become themselves highly concentrated private-profit centers, no longer serving the essential, independent function of a free press envisioned by our founders.

> One-rule economies create waste and destruction because they . . .
> 1. disrupt nature, since they can't register the damage they cause;
> 2. concentrate wealth and power;
> 3. deprive us of fair and open public conversation;
> 4. allow private power to distort public choices to serve its interests.

One result is that, for example, those with wealth and vested interest in denying climate science can use the media to shape public perception. The oil and chemical industry multi-billionaires Charles and David Koch—whose company Koch Industries ranks among America's top ten air polluters, according to a university study—have backed media campaigns scorning the scientific consensus on climate change. Surely their investment in swaying us is one reason that in just four years the share of Americans who accept that human activity is causing the climate crisis dipped from half to just one-third.[29]

## PRIVATELY HELD GOVERNMENT

In answering "Why the built-in waste and destruction?" so far I've named the root of our crisis as a "one-rule economy" that, first, is unable to register its destruction; second, inexorably concentrates wealth; and third, deprives us of an open, fair exchange of ideas.

A fourth problem may be even more serious for us all, if that's possible to imagine:

Rules that consolidate wealth also end up *distorting public decision making*; so they corrupt the very workings of democracy that we need to fix the destructive rules.

That's double trouble.

It means that in the wake of the 2008 financial collapse even corporations receiving tax-funded bailouts continued to lobby against public oversight.[30] So the financial "reform" bill passed in 2010 did little to challenge a major force behind the crash, the secretive trading of risky "derivatives"—even though we know that secrecy has proven over and over again to bring out the worst in humans.[31] And the bill left in place an even smaller number of megabanks "too big to fail": They're 20 percent larger than before the crash and control more of our economy, so if even one falters, taxpayers will again have to jump in or risk systemwide collapse.[32]

When concentrated wealth of this order infuses and distorts the political process, we end up with the ultimate oxymoron—"privately held government." The term seems apt, not flippant, when we let sink in that for each person you and I elect to represent us in Congress roughly two dozen lobbyists are at work persuading them to serve their clients' interests instead.[33] And lobbyists aren't just knocking on our representatives' doors; they're increasingly being given the keys. In the first six months following the 2010 midterm elections, nearly one hundred lawmakers "hired former lobbyists as their chiefs of staff or legislative directors," reports the *New York Times*. Congressional committees and subcommittees, trusted to draft our laws, hired forty lobbyists as staff.[34]

In 1938, Franklin Delano Roosevelt described the danger of our current predicament with startling candor:

> The liberty of a democracy is not safe if the people tolerate the growth of private power to the point where it becomes stronger than their democratic state itself. That, in its essence, is fascism.[35]

In this context, it's no surprise that public policies increasingly favor giant corporate entities and those with a direct stake in them—including tax-related policies: In 1950 almost a third of federal tax receipts came from corporations; now it's just 9 percent.[36] Everyday citizens must therefore carry more of the cost of government. The burdens of privately held government, however, go beyond a heavier share of taxes or losses from an irresponsible banking industry.

Before the disastrous blowout of BP's Deepwater Horizon drilling platform in April 2010, the giant oil company had run up 760 safety violations so serious that the Occupational Safety and Health Administration called them "egregious" and "willful." (Ranking second and third were Sunoco and ConocoPhillips: with eight and four violations.)[37] BP, it turns out, had rested the fate of human lives and entire ecosystems on a blowout preventer of a type shown to work less than half the time.[38] And, a full year later, dead baby dolphins are still washing ashore, thousands of families whose lives were ruined by the accident are still angry that they've gotten so little help, and key recommendations of the independent commission charged with protecting us against another such nightmare are still not in place.[39]

So why haven't our government's agencies tasked with overseeing offshore drilling on our behalf protected us?

The oil industry is among the largest campaign contributors and lobbying forces in Washington; and oversight of the industry has become so soft that, over five years, BP had paid just $373 million to avoid prosecution for its violations. Sound like a lot? That's less than one half of *1 percent* of BP profits during this period.[40]

Or consider the similarly powerful chemical industry's ability to fend off public oversight: "Of the more than 80,000 chemicals in use in the U.S.," editors of *Scientific American* noted in 2010, "only *five* have been either restricted or banned. Not 5 percent, five," and "the EPA has been able to force health and safety testing for only around 200."[41]

But by early 2011, the suffering of the Japanese living through the worst nuclear accident in a quarter century focused hearts and minds on one particular risk—that of nuclear power. So, a burning question for me became:

*With so many great safe-energy options, why did we risk this horrific accident, and why now are so many still willing to live with the danger that nuclear power brings?*

My rule of thumb is that when a viewpoint seems to flout common sense, it's smart to explore the assumptions hidden beneath it. In this case it seems clear that our privately held government, under the influence of the nuclear industry, has been strikingly effective in setting the frame through which many now perceive choices.

At a social gathering recently, a dear friend in his thirties, a physician and father, vigorously rejected the idea that the Fukushima accident might be the death knell of nuclear power. Nuclear power has caused a whole lot fewer deaths than coal, he argued. And renewable energy can't meet energy needs; so, despite the risks of nuclear power, he said, we can't get by without it.

We have no choice, he was telling me, and a lot of people must feel that way. After all, if we knew we had a choice, who would choose a technology in which one accident could kill hundreds of thousands? Or choose to bequeath to our children deadly Plutonium 239, just one dangerous by-product of nuclear power, which remains hazardous for as much as 500,000 years—especially when we haven't been able to agree on a single waste storage site?[42] Or choose a power source sure to heighten the risk of nuclear proliferation and to make us more vulnerable to an extremist's plot? Physicist Amory Lovins described New York's Indian Point nuclear plant as "about as fat a terrorist target as you can imagine."[43]

*So where does the frame come from that robs us of choice?*

A highly concentrated industry. Just three corporations build nuclear reactors in the US. One, GE, is the second-largest company in the world and responsible for most of the reactors in Fukushima. A few dozen more are involved in development and operation. They sure know how to spend their money. In 2010, for each of the 535 members of Congress, GE alone spent $73,000 lobbying.[44] In Barack Obama's four-year Senate career, executives of the nuclear giant Exelon were among his top ten campaign donors.[45] So it gets easier to grasp why, even after Fukushima, President Obama continued to tell us that nuclear power is safe and to push for more nuclear reactors.

The "safer and necessary" frame my friend has absorbed starts there but extends far beyond the industry's clout in Washington.

The UN-affiliated International Atomic Energy Agency (IAEA), which sounds like a neutral watchdog, says its mission is "to accelerate and enlarge the contribution of atomic energy." In 1959, at the beginning of the nuclear power era, this body and the World Health Organization (WHO) signed an agreement requiring that before any undertaking, each "shall consult the other with a view to adjusting the matter by mutual agreement."[46] Might this help explain, asks physician and nuclear opponent Helen Caldicott, why a 2009 report published by the New York Academy of Sciences predicted the death toll from the 1986 Chernobyl accident ultimately to reach nearly a million, while WHO, IAEA, and others have coauthored reports projecting a death toll one hundred times lower?[47]

Little wonder that my friend sees nuclear as safer than coal.

But the real question is why so many people feel caught between these two terrible choices in the first place. My hunch is that, in part, they believe they're just being realistic because it seems obvious that renewable energy could never gear up fast enough to meet our electricity needs.

What is missed here is that renewables already provide half as big a share of our electricity as nuclear does.[48] And it's not nuclear power, whose plants can require a decade to bring on line, but wind power that can expand quickly: US wind electricity generation jumped fivefold in just five years, from 2004 to 2009.[49] And its potential is huge: Offshore wind, by itself, according to Department of Energy data, has four times greater potential capacity than the nation's total present electric generating capacity from all sources.[50]

And surely another reason my friend sees nuclear as the only viable option, however unappealing, is that he's got no idea that he himself has been helping pay the industry's bills: Government and rate payer subsidies—totaling roughly $300 billion and still mounting—are what have made nuclear *appear* cheaper than safe, renewable alternatives, reports the Union of Concerned Scientists.[51] For almost fifty years our subsidies for nuclear power have been "more valuable than the power produced," the scientists estimate.[52]

And my friend may not know that the projected cost of building a nuclear plant has risen to $9 billion, ensuring big future rate increases, or that a federal spent-fuel repository is expected to cost American users nearly $100 billion over its lifetime.[53]

Finally, he probably assumes that nuclear power could help free us from dependency on foreign suppliers, but, unfortunately, about 80 percent of nuclear fuel is imported. [54]

## Lost opportunities

Luckily, I later had a chance to go over some of these points with my friend. "I clearly I have a lot to learn," he reflected, but then asked great questions about the viability of a renewable energy future and the best ways to get there. Unfortunately, quickly finding the best answers to these questions gets thwarted by exactly the same forces ensnaring us in fossil fuels and nuclear. From 2002 through 2007, for example, well over six times more public monies went to fossil fuel- and nuclear electricity–related research and development—over $9 billion—than to renewables.[55] Imagine where we'd be if that $300 billion in nuclear subsidies over the last decades had gone to research and development of renewable energy.

Or consider this missed opportunity:

When the government injected significant stimulus money into our economy to stave off depression in the aftermath of the financial crisis, it sure seemed like the perfect opportunity to shift course to energy sanity.

But what happened?

The "green" portion of the 2009 stimulus package came to a measly 12 percent. Think where we'd be if we had taken a course more like that of the EU—where 59 percent of government stimulus funds went to green initiatives. Or better yet, like South Korea at 80 percent.[56]

I mention these missed opportunities to underscore that our problem isn't the politics of a particular misguided administration, or a greedy few at the top, or our society's fixation on growth. It is rather a deep "design flaw" in our particular version of the market—one putting private over public interest—that is killing us.

## REFOCUSING ON WHAT MATTERS

So here we are. And calls for "no-growth" clearly can't save us.

Distinguished economist Herman Daly sees it differently, advocating "no-growth" and "steady-state" economies because growth always involves "quantities" that are "basically physical" and therefore fill up the planet. Yet his goal, and mine, is what he calls "qualitative development," that is, what enhances the quality of our lives and our ecosystems.[57] To get there requires, I believe, that we leave behind metaphors that keep us focused on additive *quantities*—too much stuff—and seek those that encourage us to probe the destructive *processes* that create vast deprivation, even amid plenty; processes created by centralized control systems that are designed with no means to register their destruction.

Plus, a framing that can be received as focused on quantities of things fails to encourage us to see that many "things" of value for which we exchange money aren't "basically physical." Our motivation can quicken as we appreciate that only our imagination limits the number of ways we might enhance each other's lives and get paid for it. Think art, sports for fun, child and elder care, religious activities, healing, environmental restoration, and education and lifelong learning.

## BEYOND GROWTH VERSUS NO-GROWTH

As we learn to think like an ecosystem, we can see a way forward: Realizing that ever-evolving relationships define life forms and experience—including ours—we can let go of mechanical dichotomies, like growth versus no-growth, empty or full. Finding our courage to reject privately held government, we can shape economies that register and respond to nature's laws in a democracy accountable to the citizens.

> Economic development isn't a matter of imitating nature. Rather, economic development is a matter of using the same universal principles the rest of nature uses.
>
> —Jane Jacobs,
> *The Nature of Economies*, 2001

Then, in economies harmonizing the meeting of real human needs with nature's generation and regeneration, entirely different questions come to the fore. Not more or less growth, but questions like these:

What does wider nature teach us about the human social environment?

How can we create rules for our societies that take their cues from nature's rules—rules that keep energy, of which money is one medium, moving in ways that enhance life, instead of trapped at the top of an economic pyramid? For unless money is circulating widely, none of the promising new green economic initiatives can truly bear fruit.

Strategies that keep wealth dispersed and fluid are in many ways pretty straightforward. They include such prosaic ones as taxing households according to ability to pay—as Adam Smith, a favorite of conservatives, recommended more than two hundred years ago—and a minimum wage high enough that the "free market" comes to mean one so open and fair that we are all free to participate in it.[58]

Equally useful to keeping wealth circulating, so we're all free to be participants in the market, are rules—abandoned here thirty years ago—that bar interest rates so high that they trap borrowers in perpetual debt. Also needed to keep market access open and money flowing throughout an economy are rules that prevent corporate monopolies and break them up when needed, plus rules that ensure workers' freedom to join labor unions without fear of reprisal. Access to education and health care is essential as well, so that each of us can be a contributing community member, both creating and using wealth.

Such straightforward approaches depend on government answering to citizens' interests and, therefore, on removing the grip of concentrated wealth on public choices—which I'll take up in Thought 7.

We've done it before—from the 1930s to the 1970s, the US made significant strides on these fronts—and around the world today, some companies, communities, states, and whole societies are already breaking with the growth versus no-growth paradigm in order to thrive.

Here is a mere taste of these breakthroughs holding huge lessons for us all.

## The flourishing company

I'm sure the late Ray Anderson never imagined himself becoming an icon of ecological enterprise. But years ago, Anderson, founder of Interface Inc., experienced a personal epiphany—his own moment of dissonance: Reading Paul Hawken's *The Ecology of Commerce*, he suddenly realized the environmental impact of his carpet company. Anderson soon stepped out ahead of the pack, and in just the last seventeen years, his global carpet giant has made huge strides.

The company's success in part flows from a breakthrough in perception. Anderson stopped viewing a carpet simply as one big, continuous surface and introduced carpet "tiles," which allow replacement and recycling of specific worn-down or soiled modules without wastefully disturbing the rest of a perfectly good carpet.

Interface asked designers to spend a day in the forest and to create carpets inspired by the forest floor. The designers realized "'there are no two things alike here.' No two sticks, no two stones, no two leaves, anything," Anderson remembers. "Yet there's a pleasant orderliness in this chaos." When they came back, they designed the carpet tile they call "Entropy," which became the company's biggest seller.[59]

And randomness turned out to offer a real ecological and efficiency dividend. Because of its random pattern, Entropy involves less installation waste. Installers don't have to worry about a tile's direction. Random design also means that when worn or stained tiles need replacing, new ones blend in easily.[60]

Interface has pushed to reduce its greenhouse gas emissions partly by becoming more energy efficient, partly by turning to renewable energy, and partly by making carpets with one-third to one-half less oil-based yarn than competitors'. It has captured methane emitted from a landfill to power one of its plants. Interface has cut emissions by almost half and reduced water use by 80 percent.[61] Anderson described the mountain Interface is climbing as "taller than Mount Everest—it's Mount Everest Sustainability." And at the top is "Zero Environmental Impact . . . where we really want to get to."[62]

Along the way, Interface's sales have grown by two-thirds.[63]

Today, the company's goal, prominently announced on its website, is to eliminate all negative environmental impact of its operations by 2020.

## The flourishing farm community

Far away from corporate life and carpets, imagine farmers in India scared to death—literally—when their crops fail, and they're unable to repay debts taken on to buy "modern" corporate-controlled seeds and chemicals for crops destined for unstable global markets. Imagine the stress and illness as they struggle from dawn to dusk. The southern state of Andhra Pradesh has long been a center of such misery dubbed the "pesticide capital of the world" and among those Indian states hit hardest by farmer suicide.[64]

But, in the 1990s, some farmers began to see their entrapment with new eyes.

With loans from civil society groups and then support from the minister of agriculture, what locals call a "nonpesticide" movement began to spread in Andhra Pradesh. A number of cotton farmers began to say no to Monsanto Corporation's genetically modified seeds and returned to local varieties, costing a fraction of what they had paid for the patented corporate seeds, and to natural pest-control practices. Their costs fell, and soon profits began to rise. I was struck by a photo in the journal *Seedling* of a poster in the center of a village displaying, for all to see, the outcomes of the two farming options: Farmers using the nonpesticide approach enjoyed 23 percent more net income than their chemically dependent neighbors.[65]

As the movement has spread, farmer indebtedness has fallen drastically, and a survey in three Andhra Pradesh districts showed hospitalizations from pesticide poisoning had dropped 40 percent. The local economy quickened as former pesticide pushers, who previously returned profits to corporate producers far away, began making and selling natural pest-control potions using the ancient neem tree, cow dung, chilies, and other locally available ingredients, keeping profits in the villages.[66]

Touching only 400 acres seven years ago, by 2009 sustainable farming practices had reached roughly 1.3 million acres spread across most of the state's districts—with no sign of slowing down.[67]

At the same time, in a related dramatic step, poor, low-caste women in a handful of villages in the state's Medak district came to realize another kind of dependency, beyond that on high-cost seeds and toxic chemicals: Many had abandoned their fields of nutritious, diverse millets, lentils, amaranth, and greens and begun to rely heavily instead on eating government-subsidized, chemically produced white rice—lacking essential nutrients. Ah, they realized, no wonder their health and communities were suffering.

Together, the women's groups, working with the local Deccan Development Society (DDS), found their courage to reclaim abandoned fields as well as traditional knowledge of ecological farming practices, including the saving and sharing of seeds. With their healthy harvests, women consistently repaid loans received through the DDS for land reclamation.

They also devised their own food-security reserve to end hunger and indebtedness: It begins with an outdoor meeting in which a whole village comes together to decide which families will likely need extra food between harvests, like those with little land but many children or a widow's family. Then all households contribute to a reserve from which those with extra needs can draw.[68]

This women's network has now spread to involve almost 6,000 households in 125 villages. They've calculated that their work is already producing almost 3 million extra meals each year, as well as almost 350,000 additional days of employment. And it's spawned a yearly festival celebrating seed saving and sharing. At first only a few joined in, but now, thousands—with music, dancing, art, and a colorful seed caravan traveling from village to village.[69]

The women of DDS, mostly illiterate, have not only found their voices but are broadcasting them: They've learned to produce high-quality video and community radio to document and share their successes.[70]

On very different terrain in northern India's Himalayan foothills, my daughter, Anna, and I hiked into villages a decade ago to meet those in a similar learning-doing network—Navdanya—founded by one of the world's most effective champions of ecological farming, Indian physicist Vandana Shiva. The Navdanya farmers we met, who not long before had been hooked on commercial seeds and on pesticides, glowed with pride as they showed us their handmade registries of indigenous plants they are protecting through ecological practices. Since our visit, Navdanya has not only trained 500,000 farmers to make this transition to eco-friendly farming but has created its own farmer learning and field research center near Dehra Dun, conserving 3,000 varieties of rice.[71]

All this is indeed growth—stunning growth of an ecologically sound farming system and community health.

## The flourishing local economy

Breaking with the assumptions of a "one-rule" economy that is centralizing control, citizens worldwide are beginning to assert their power to stand up for their local economies, decentralizing power.

In the US and Canada, in eighty towns and cities, a "Local First" movement of independent, sustainability-oriented businesses has taken off.

Together they're encouraging purchasing that keeps the wealth of their customer neighbors near home.[72] Bellingham, Washington, took an early lead in this love-your-hometown movement.

"Are you locally owned?"—more and more businesses are hearing the question, Michelle Long, founding executive director of Bellingham's Sustainable Connections, told me. "Before, the question never came up," she said.

Now, more than 650 participating independent businesses in this city of 80,000 display a poster and a "buy local" decal in their windows. They give special thank-you cards to loyal customers and offer customers coupon books with discounts at member stores. To kick off the campaign, Sustainable Connections welcomed citizens to compete to collect the most receipts from local businesses in one month. The grand prize? A month of free meals at locally owned restaurants.

A dollar spent in a locally owned business can generate three times more local economic activity than a dollar paid to a corporate chain.[73] Not to mention the community building that happens when we shop from proprietors we know and the enormous environmental benefit of shortening supply chains.

In Europe, a superhero of this awakening to the possibility of thriving local economies is the small eastern Austrian town of Güssing, which only twenty years ago was one of the poorest towns in a poor region. Seventy percent of its workers had to commute to find work; millions of dollars were leaving town every year to buy fossil fuel. But local leadership began to see possibility: Why not, they thought, stop the outflow by becoming energy efficient and by taking advantage of what they had in abundance—woodlands, potentially offering cleaner energy—all while investing in local jobs?

Soon, ambitious conservation measures in buildings in the town's center had cut energy use in half. Today, mainly through an innovative biomass (wood chips plus agricultural waste and sewage) gasification process, Güssing is now able to supply all its own energy—with virtually zero emissions. It even has enough to export some too. The abundance of clean energy has attracted fifty new businesses to Güssing, creating 1,000 new jobs.[74]

And there's another, unexpected income stream: For as long as townspeople could recall, its twelfth-century castle had been Güssing's only real tourist attraction, but tens of thousands of tourists now make their way there each year to be inspired by Güssing's green turnaround.

This global movement is not just greening physical power; it's also dispersing social power.

Worldwide, the movement includes 800 million members of cooperatives—enterprises owned by their users, including shoppers, savers, farmers, and small business owners and workers, not by distant investors and speculators. Membership has jumped tenfold in the last half century, "making lives more secure for almost half the world's people," estimates Cooperative Hall of Fame honoree David Thompson.[75]

And perhaps nowhere more dramatically than in the Emilia Romagna region of northern Italy: There, over 5,000 cooperatives make everything from the country's best parmesan cheese to its first electronic scale. About a third are small worker-owned companies with fifty or fewer members, while some consumer cooperatives boast 1 million plus members. Together they generate at least a third of the region's economy.

Visiting Italy a few years ago, my guide Davide Pieri, who worked within the co-op movement, told me he'd tried working for a corporation and then as a private consultant, but, he smiled, "This is the interpretation of life that I enjoy." It's a more democratic economic model that has helped make the area one of Europe's most prosperous.[76]

At the other end of the prosperity scale is Cleveland, Ohio, recently ranking number one on the Forbes Misery Index. There, ten major cooperative enterprises are getting under way to move misery to opportunity. One, the Evergreen Cooperative Laundry—industrial size and thoroughly "green"—opened in 2009 to serve health-care facilities in a neighborhood whose median household income is only about $18,000. Each of the planned worker-owned businesses, projected to offer a total of five hundred jobs, will contribute 10 percent of pretax profits into the seed fund to help launch still more co-ops, writes Gar Alperowitz in *The Nation*.[77]

Elsewhere, citizens are stepping up as joint owners of enterprises their communities depend on.

In the center of Powell, Wyoming, population 5,500, for example, stands the Mercantile, a 12,000-square-foot department store run not by a multinational based in Arkansas but by its owners—all five hundred of them—who live within blocks of the store. The idea for "the Merc" was born in 2001, after the town's only general clothing store, owned by an out-of-town chain, closed. Suddenly, even buying a pair of shoes required driving twenty-three miles to Cody or one hundred miles to Billings.

Inspired by what the people of Plentywood, Montana, had done in a similar pickle a couple of years earlier, the Merc's volunteer committee needed only a few months to sell eight hundred shares at $500 a pop—enough to open the doors. In the first year, the Merc's gross sales shot well past its board's projections. By early 2004, the store had outgrown its space and expanded into a 2,500-square-foot basement nearby.

The Merc continued to succeed even after a Walmart Supercenter opened only forty-five minutes away. And in 2010, Merc shareholders began to enjoy dividends on their investment.[78]

Now folks from other towns in the heartland arrive in busloads to learn from the Merc. And elsewhere, the movement is growing, as in the UK, where over two hundred towns now enjoy thriving community-run stores.[79]

Also contributing to thriving local economies is the takeoff of local and regional complementary currencies. Not backed by the national government, they're traded only in a defined area for services and locally produced goods—with the goal of keeping wealth circulating locally.

In Germany, you'll find hundreds of complementary currency systems, roughly thirty of which serve whole regions. In 2003 in the Lake Chiemsee region of Bavaria, Waldorf School economics teacher Christian Gelleri and his students created the Chiemgauer note (equal in value to a euro). Since then, some 3 million have begun circulating, with six hundred businesses now accepting them. A regional currency like the Chiemgauer is spent on average eighteen times in local commerce—that's three times the local circulation of the euro.[80]

In the US, western Massachusetts's BerkShares currency (each note is worth $1 and purchased for $0.95) has grown remarkably. Since 2006, $2.4 million worth of BerkShares have flowed into circulation, accepted by more than four hundred businesses. Cofounder Susan Witt explains one way BerkShares build community wealth: "Businesses are now trading with other local businesses, so that they're sourcing their printing, accounting, and food products locally rather than out of the area."[81]

## The flourishing country

Over several decades, Costa Rica—with a per capita income just one quarter that of the United States—has made startling advances despite marked income equality.[82] Health care and education have expanded to make people's lives longer and more fulfilling. In just one decade, the 1970s, the country slashed its infant death rate by more than two-thirds. By the 1980s, 60 percent of households benefited from quarterly health worker visits.[83] In life expectancy, Costa Rica ranks just behind the US and Canada in a virtual tie for third place.[84]

Three decades ago, Costa Rica had one of the world's *worst* rates of deforestation. Forests had shrunk to cover only one-fifth of the nation's land. But over the last two decades, the country has used smart incentives to encourage farmers to protect trees, and now forest covers about half the country.[85] Even more surprising to some, recall Costa Rica's decision to place a moratorium on oil drilling.[86]

Encouraging green energy sources, Costa Rica has achieved an ecological footprint per person just one-quarter that of the US. By 2009, Costa Rica had risen to first place in the world on the Happy Planet Index, ranked by its citizens' perceived well-being and the size of a country's "ecological footprint."[87] In Thought 7, I'll explore why Costa Rica's path is different from that of so many of its neighbors.

Each of these far-flung examples—which many of those involved would certainly describe as "growth"—reflects life-enhancing economics that are improving the lives of citizens while reducing the built-in system of waste I decried earlier.

## MEASURING OUR PROGRESS WITH AN ECO-MIND

Such stories can spur us to get over the useless growth-versus-no-growth debate. But to truly succeed, we need better ways to measure what we value as Herman Daly pointed out two decades ago. And helping us do just that is a burgeoning international movement radically rethinking the gross domestic product (GDP)—that grossly misleading measure of a society's well-being that is simply the sum of expenditures for goods, services, and exports minus imports.

To grasp the GDP's shortcomings, in just one example, recall the earlier tale of indebted Indian farmers. Then consider that in 2010 India's GDP got a boost when the State Bank of India provided to a businessman in Maharashtra—the state ranking highest in farmer suicides—credit to buy 150 Mercedes Benz. The GDP grows with one man's frivolousness. But suicides of Indian farmers, ruined by lack of fair credit and prices, occur in India every thirty minutes, on average. Yet the GDP subtracts nothing for their demise.[88]

A very different measure, tracking well-being from a relational perspective, is the Genuine Progress Indicator (GPI), brainchild of Ted Halstead, founder of Oakland-based Redefining Progress. The GPI looks at both how much money we spend *and* what we gain or lose with that spending.

The cost of crime—measured in money spent to replace damaged property or on medical and legal expenses—pushes up the GDP, but in the GPI this cost is deducted from the measure of national well-being. The GDP adds the value of natural resources we extract and sell, but the GPI looks at the true costs of that use—subtracting, for example, the $1.18 trillion cost of carbon emissions in just one year, as well as the destruction of forests and farmland.[89]

So, while our GDP has grown steadily, our genuine progress—measured not only in financial wealth but also in the positive value of leisure time and fair income distribution and the cost of environmental damage, to name a few aspects—is hardly better than it was in the mid-1970s.[90]

The GPI encourages us to drop the narrow "more-or-less" lens and to ask:

Are our societies flourishing, and how do we know?

What does any quantitative measure tell us about the *qualities of the society* we want to grow?

Using a GPI, it's likely we'd make different choices. Here's a glimpse of what that might look like: For three years, citizen protests against building a third runway at London's Heathrow Airport were stymied. But in 2010 the protestors prevailed after an independent study took a wider view. It considered costs of additional carbon, congestion, and pollution, for example, and the project's touted economic gain turned into a cost to the UK "upward of £5 billion."[91]

Polls show that majorities in many countries favor more accurate measures of well-being, like the GPI.[92] The "good news," says Italy's chief of national statistics, Enrico Giovannini, is that in half a dozen countries, including Germany, the United Kingdom, and France, top officials are expanding their focus beyond mere quantitative economic growth to "well-being." Even China has embraced fifty-five green indicators.[93]

Seem unlikely to find favor here?

I would have thought so, but in early 2010, Maryland became the first US state to develop and adopt the GPI to measure its progress, using twenty-six indicators of economic, social, and environmental well-being.[94]

The state intends to use a lens that captures well-being to plan for the future, so it turned to the University of Maryland's Center for Integrative Environmental Research to develop a modeling tool to forecast how investments today would affect prosperity fifty years from now—measured by the GPI. It projects that going all out for green jobs, clean-energy savings, or what the center's director, Matthias Ruth, calls "smart growth," would each double the state's GPI.[95]

Certainly, many people hearing about the breakthroughs in this chapter would perceive them as examples of growth, really positive growth. Maryland splashes these words across its website promoting its new measure of progress: "Smart, Green and Growing." Surely, "growth" is a word I personally don't want to give up. I want my tomatoes to grow, and my friendships.

But the word, applied to societies and economies, has for so long meant only a quantitative measure of expansion that it's probably time to give it a rest. To communicate the enhancement of qualities we want, rather than the expansion of quantities we probably don't, we might be a lot more effective using other phrases and terms: progress toward well-being, for example, as in the Genuine Progress Indicator. Or we might choose, as I have here, to speak of "flourishing," "thriving," "life-enhancing" economies and communities.

And why, you might wonder, not simply stick with the much more common term "sustainable"? It's all right. But "to sustain" suggests "bearing up" or "keeping on," and I want more. And I think most of us do too.

## A CLARIFICATION

Before moving on, let me be super clear.

In challenging the "growth-versus-no-growth" frame, I am in no way putting myself in the camp of analysts such as Julian Simon, author of *The Ultimate Resource*, who have argued for decades the naïve notion that as we continue to practice business as usual, market signals, combined with infinite human creativity and a desire for profit, can meet any environmental challenge. Both this faith-based, market doctrine and the limits-to-growth frame share the same premise: scarcity. One says growth is the answer to scarcity; the other suggests that growth is the central problem creating it.

I'm saying neither. I agree strongly that today's economic "growth" is *not* working, but to define what we've been doing as "growth" risks blessing our current practices with a term that sounds positive to most ears. That's a problem. Plus, "no-growth" can look downright scary to the jobless, who understandably see economic growth as essential to putting bread on their tables.

Most troubling to me, however, is that the focus on growth as the problem—or as the solution—keeps us from probing to the root of our global crises: the patterns of power over decision making that we ourselves choose, consciously or not, that leave one kind of deadly economic model gaining in strength. Instead, I'm suggesting that we can utterly shift our vision to the goal of aligning our practices with nature, including all we now know about human nature. Together we can then get on with creating the context—the social and ecological relationships—that enable all of us to flourish, to experience a plentitude of what really matters to live fulfilling lives.

## THOUGHT LEAP 1

Since what we've been calling "growth" leads to so much waste, let's just call it what it is—an economics of waste and destruction. Then let's probe *why*. Dropping the distracting "growth-versus-no-growth" debate, we can embrace qualitative notions of where we want to go, choosing terms like "flourishing" and "genuine progress" that focus our minds on enhancing health, happiness, ecological vitality, resiliency, and the dispersion of social power.

Closely tied to the "no-growth" thought trap are two much more widely heard framings of the problem: one, that we consumers are to blame for the mess, and, two, that humanity has hit the limits of what earth can sustain. Let's look at each of these thought traps in turn.

## thought trap 2:

# "Consumer Society" Is the Problem

*Out-of-control shopping is overtaxing natural resources.*

T HE DAY I BEGAN THIS CHAPTER, NPR'S *Marketplace* FEATURED
"designer diapers," a baby's first fashion statement. Apparently,
their premium price had not stopped even Walmart shoppers from snap-
ping them up.[1]

Just one hint of the ravages of this kind of shopping mania is this stunning estimate: On top of all the stuff itself, every year on average each
American now disposes of almost three hundred pounds of packaging,
and that's without even counting a lot of soda and beer cans.[2] Visualize it:
almost a pound a day for every man, woman, and child in the country! No
wonder Americans generate about a third of the world's trash, about a ton
per person each year.[3]

So it sure is easy to join the chorus that "consumerism" is out of control, and it is killing us.

"We in the industrialized countries use way too much stuff," says Richard Heinberg, author of *Powerdown* and *The Party's Over*, "because that stuff is made from depleting natural resources." Heinberg notes that every environmentalist rails against the "dragon" of consumerism, which must be slain.[4] The US is the world's "champion consumer," declare veteran environmental scientists and advocates Dr. Paul R. Ehrlich and Anne H. Ehrlich in *The Dominant Animal*, because of Americans' "disproportionate consumption and attendant environmental destruction."[5] From this angle, the problem appears to be "consumer society," and the diagnosis sounds like common sense.

But "consumer society is to blame," as a central environmental message, can't work. Even the word "consumer" misleads, since none of us consumes *anything*. Whether it's food or a laptop, we are mere "passthroughs," and all that passes through us goes somewhere. We may think we throw things away, but there is no such place as "away" in an ecological system.

Before really digging into why I believe this approach can't work, however, I want to be sure I'm being fair to environmentalists focusing on consumerism. Many don't simply blame shoppers; they also condemn corporations for aggressively stimulating "consumer demand" and, like Paul and Anne Ehrlich, also implicate the concentration of power. But my focus is on what gets *heard*, and very often it's simply that overconsumption in the "rich countries" has created the crisis, period.

That message is a nonstarter for a couple of reasons.

For one, it doesn't fit most people's everyday lives, even in an ostensibly wealthy country like the US. True, the stereotypic "American consumer" is the guy lugging a flat-screen TV through the doors of Walmart on Black Friday. But what does this image hide?

Most of the percentage increase in "consumer spending" over the last thirty years went to pay not for new TVs and the like but to escalating health-care costs. In fact, the share of the US gross domestic product devoted to spending on goods fell almost 5 percentage points between 1978 and 2008.[6] And that was before many of the staggering losses Americans have more recently felt: Between spring of 2007 and late 2009, the net worth of American households fell more than a quarter.[7]

Plus, if we see our problem through the lens of consumerism, it's easy to feel that it's all somehow inevitable. British environmental leader Tim Jackson writes that "our own relentless search for novelty and social status locks us into an iron cage of consumerism."[8] And since both a passion for novelty and a need for social standing seem to be deeply human traits, I might easily misunderstand Jackson to be saying that consumerism reflects innate impulses, even though his own analysis is much more nuanced. How depressing, and de-motivating. Then, on reading John Naish, author of *Enough: Breaking Free of the World of More*, lamenting our "limitless instinct to pursue more of everything that we enjoy," my heart sinks further.[9]

Closely related to the downer of being told that consumerism is virtually a genetic given, this thought trap seems certain to make people in the Global North feel accused. Yet, whether learned the hard way as a mother or as a world-hunger fighter, I've come to see guilt as a pretty lousy motivator. Ultimately, people don't like having the finger of blame pointed their way, especially when they feel they're struggling just to get by themselves.

Plus, if this message is heard as a call to "give up" the very goodies that make increasingly stressful lives a bit more fun, convenient, and comfortable, then of course those hearing it would assume that any *other* way of living must be dull and trying.

## Born to buy?

For starters, we gain ground toward solutions by distinguishing between what is innate in all this—humans' love of novelty and our need for social approval, for example—and what isn't. True, these proclivities make us *vulnerable* to manipulation to buy ever more, but that's a far cry from assuming that our vulnerability makes endless accumulation almost inevitable.

Pondering how much of today's material fixation is "natural," I sometimes return to a surprise I got years ago reading my own grandmother's (three volume!) very personal history of our family.

Born on the Kansas prairie, Frances (for whom I'm named) married and gave birth to my dad at an age that was then considered late—

thirty-seven. Her husband, George Moore, ran the feed store in the small town of Winfield. I loved reading the rich details of her daily life, but what stood out most was that virtually none of her preoccupations had much to do with *things*. Her very full life revolved around church, piano playing, friendships, the boys' club she started, and other community places and events—like the county fair in a park where she met my granddad.

I mention my grandmother's life because its meaning is often overlooked. For me, it's a reminder that the way we, the better off in the Global North, now live—focused so centrally on the material—is *really* new in our species' evolution. It is an aberration, as I've argued earlier. It's not an inevitable expression of humanity.

Yes, maybe, I can hear a reader pushing back: "But modern life has so many advantages that few would ever *choose* to give them up and live like your granny. Isn't consumerism therefore *effectively* inevitable?"

This response, I believe, carries a confusing and false assumption: that "consumerism" is just another way of saying "creature comforts," convenience, and excitement. So when one imagines my grandmother's life, no matter how appealing its more relational aspects, we wouldn't want it. God forbid. We assume we'd surely be physically uncomfortable and both intellectually and emotionally constrained. In a word—life would be really *dull*.

We have difficulty imagining lives that are both relationally rich, unburdened by the stress of amassing and protecting a lot of material things, *and* also replete with creature comforts (pleasant indoor temperatures and hot baths), enjoyment of novelty (the iPad or exotic travel), and intellectual and emotional stimulation.

In a word, we conflate modernity and consumerism. But we can get over it. We can begin to imagine how modern societies, aligned with the laws of nature—and, yes, reflecting human nature—could be vastly richer than the stressed lives so many lead today.

## MATERIALISM MISDIAGNOSED?

To get started, we might first question whether at least part of what's assumed to be proof of "materialism" is in fact a reaction to our inability to secure what we really want. In other words, when societies' governing rules and expectations deny human beings ways to express our deep needs and capacities for security, fairness, cooperation, and efficacy, which I take up in Thought 4, we don't just fold our hands in defeat. Sometimes we go for the next best thing—trying to satisfy some of what's missing through our badges of success, our stuff.

So here's the irony: Much of what gets labeled self-seeking consumerism may actually reflect our deep need for connection with others. The seemingly endless purchasing by some modern humans might not be about "things" at all or the comfort or convenience they bring us. Might the urge—be it for a McMansion or cool new sneakers—really be about relationships, our yearning to enhance our status with others?

Fascinating research reveals that merely touching money—not even being given it—can ease the pain of social ostracism (as well as diminish the sensation of physical pain). When asked about how touching the money affected them, subjects surprised researchers by noting that it made them feel stronger.[10] And, as I've argued, our need for power runs deep.

Buying things seems also to be a common human strategy to ease feelings of insecurity and fear. When we humans are reminded of our own mortality, we focus more on "money, image and status," social scientists Tom Crompton and Tim Kasser report.[11]

If there's truth here, then an effective strategy for calming our mad shopping may not be to scold us as selfish consumers. Might the scorn only diminish further our sense of self—and therefore backfire? If so, much more effective would be work to foster new, compelling connections in communities of common purpose instead of common purchases—a shift enabling us to feel more secure and powerful.

The burgeoning local, organic foods movement might be seen as one beautiful pathway. Through it, neighbors are creating school and community gardens and supporting farmers' markets, not merely for the physical pleasures, a plenty, but with a sense that they're part of something big and important: a fast-growing, global movement for healthy growing and eating that ties us even to the faraway Indian farmers described earlier, those rediscovering their own rich food heritage.

I'll never forget a conversation, years ago, with the late Pauline Thompson, an elder member of Kentuckians for the Commonwealth, a powerful citizen group that had successfully taken on the state's dominant coal industry. She told me that she used to feel that if her neighbor bought a new couch, say, she had to run out and buy one too. But once her community work captivated her, she found she had no patience for such trivial matters.

With this clarity, we can challenge the common claim that, within capitalism, firms automatically produce the goods and services people need and desire. We can question the phrasing of countless think tank projections citing "market demand" as the force behind massive and rising resource overuse—as if "market demand" were an inexorable force existing on its own, apart from us.

Dangerously misleading.

Instead, we can show how corporations operating within a "one-rule economy," defined in the previous chapter, so powerfully manipulate us; and beyond that, we can focus on what assumptions and rules are giving them their power and what we can do to ensure that our economies serve life.

## INSIDE "MARKET DEMAND"

"Market demand"—what people are willing to spend—is largely a function of the rules we together are creating, allowing corporations to control so much wealth that they can spend globally each year about half a trillion dollars on advertising, *about $70 for every human on earth*, to entice us to buy what brings them greatest return.[12]

In the US, make that $1,000 per person.[13]

Since the billions thus spent on advertising end up in the prices we pay, you and I each day are footing the bill to be seduced. That we've permitted advertising to be a corporation's tax-deductible expense means we're also paying in lost public revenue. We are, in effect, paying to construct an information system using the most powerful persuasion tool known to humankind, *fear*: fear of attack ("buy my home alarm system"), fear of social isolation ("buy my wrinkle-erasing cream"), or fear of being one-down ("buy my newest iPhone before the next guy").

Also, what's lost in the assumption that "market demand" is the root of environmental problems is the significance of its *composition*: that is, who's making the demand and for what and why. Here's what I mean: In a one-rule economy, wealth inexorably moves into fewer and fewer hands, so that more and more "demand" comes from fewer and fewer people.

So, for example, while most Americans still imagine ours to be a "middle-class" society, it isn't. In 2010 the Obama administration placed households with up to $250,000 annual income in this category. *Middle?* Not really. Just 2 percent of US households are above this income mark.[14] But the most dramatic measure of concentration in the US is that the top 1 percent of households holds more net worth than the "bottom 90 percent of households put together," according to Citigroup.[15] In fact, economic inequality here, perhaps the greatest in our history, is more extreme than in India, where in 2009 the number of Indian billionaires doubled, while nearly half of all the country's children suffer malnutrition.[16]

The wealthy stimulate "market demand" of a certain kind—for big cars, airplanes, multiple homes, technology, grain-fed meat, seafood, and imported luxuries—all with consequences for the environment. If the same total dollar value of demand, in India, for example, were instead coming from millions of poor villagers creating thriving local economies tapping renewable resources—as is already emerging in parts of this country and beyond—the eco-implications could be vastly more benign.

Another ecologically consequential feature of a power-concentrating economy like ours is that workers have to put in longer and longer hours to make ends meet. On average, Americans spend roughly four hundred more hours each year on the job than do the Germans or Dutch. Our long hours warp market demand in ways damaging the environment.

"Time-poor" people tend to eat out more, use more carbon-intensive travel, and have less time for pro-environment activities like gardening or do-it-yourself projects involving fewer purchases. So the Center for Economic and Policy Research estimates that US energy use could drop by a fifth if Americans simply enjoyed the time-use patterns of western Europeans. Note that fifty years ago, we did enjoy shorter working hours than they.[17]

Thus, when you hear even distinguished environmental analyst Lester Brown argue that rising prices for grain, and thus grain-fed meat, are the consequence of "rising affluence" among the world's bottom 3 billion, don't buy it.[18] Affluence is rising for *some*. The pressure largely reflects deepening inequalities, giving a minority greatly increased market power while leaving multitudes without work and driving many of those working to put in longer hours.

Most often, this "affluence" explanation for food-price hikes and food-related environmental pressure is presented as a virtual inevitability. It's not.

Staying with meat for a moment: Note that in developing countries, animal-food consumption from grain-fed livestock that can push up prices and cause environmental damage is highly concentrated. Well over half of that increased consumption is happening in China, and to a lesser degree in Latin America, largely serving better-off urbanites.[19]

By contrast, if rural families throughout developing countries—who make up 70 percent of the world's very poor—were becoming more "affluent" and eating more animal food, the impact could be quite different. Instead of feeding grain to animals in large commercial feedlots, small family farms could be integrating livestock into productive ecological farming.[20] They wouldn't be adding pressure on food prices, and their affluence could make positive environmental ripples.

So the composition of "market demand," not just its size, matters enormously to our ecological well-being, and it isn't an inexorable force before which we are simply observers and victims.

## What is luxury?

Moving beyond the false dichotomy—either material fixation or boring lives—starts with rethinking luxury itself. It happened to me unexpectedly.

In 2006 I journeyed with my extended family—ages six to seventy-three—to the Amazonian region of Peru, eventually moving in a small boat along the Tambopata River.

After several hours, we hopped off the boat for a short hike on a narrow jungle path until we reached a large clearing. There, we got our first glimpse of our resting place, an airy structure built largely of bamboo and on stilts. Climbing the wide stairs, we entered a large, open, multilayered gathering space with high ceilings, flowers, and decorative wall hangings. I was enchanted.

And that was before our first meal in a dining room with colorful macaws swooping in to alight on high wood beams above us. And before I strolled in the early evening stillness along an open "bridge," lit on both sides by lanterns, to my bedroom, where my mosquito netting appeared more like an elegant canopy than an insect barrier. And before I lay listening to the soft jungle sounds, so clear and present because the room's outer wall reached only to my waist.

[ Luxury is beauty, but beauty is not a luxury. ]

We stayed just a few days. But as we packed into the same small boat to leave, I realized my perception had been forever altered. I had learned that, for me, *luxury is beauty* but beauty is not a luxury.

This hotel had little electricity, no private baths, and no chandeliers, no carpets, no spa, no granite fountains. But it was by far the most luxurious hotel I'd ever stayed in. My need for creature comfort was more than met—it was indulged, as I savored local dishes and rocked in the colorful hammock watching the monkeys play. And my mind was stimulated delightfully by encounters with scientists studying the rain forest and my own observations of the flora and fauna along the jungle paths.

My epiphany, redefining luxury as beauty, happened far from home through a particular experience I realize that, because of its location, isn't available to most of us. Yet, much of my pleasure came from being intimate with the natural world, and that opportunity *is* available to virtually all of us. Thus, opportunity to rethink luxury is everywhere, at any moment—including right now in my Boston home as I sit here taking in the beauty of delicate icicles, dangling at least four feet just outside my dining room window.

"Luxury as beauty" has nothing to do with a particular place or an object's price tag. It is seeing with eyes for beauty. Once we cut the automatic but learned connection between buying stuff and pleasure, we can actively cultivate new connections—a sense of freedom as we shed draining habits and discover new pleasures in seeing and creating beauty all around us.

## OUT OF THE CAFETERIA LINE INTO THE KITCHEN

Another way to grasp the shortcoming within "consumerism itself is the problem" is to reflect on certain changes that have contributed big-time to ecological disruption over the last sixty years but have nothing to do with buying more "stuff" per se. So, they can't be addressed by a simple "buy less" strategy.

Let's take just three.

One is the expanding size of homes, using vastly more resources to build and power. On average our homes are more than twice as big as they were in 1950—that means almost 1,400 more square feet to heat and care for.[21]

Second is the greater distance we and our goods travel. Commuting time increased by nearly 20 percent between 1980 and 2000, and commuting distance jumped by over 40 percent from the mid-1980s to the mid-2000s.[22] (And in most big urban areas, our time spent stuck in traffic—now equal to a full workweek—jumped threefold in twenty years.)[23] Produce also travels more than 20 percent further.[24]

Third is the change in our choice in meat. Over the last half century, Americans have shifted from range-fed to feedlot beef, using vastly more resources per bite.

So, why has all this happened?

Did families sit around kitchen tables in the 1950s and 1960s saying, "Hey, we all want to commute almost an hour a day, and we're dying for kiwis from New Zealand? And, by the way, life would be so much better if our steaks and hamburgers were fattier."

Probably not.

And even our much larger homes? Are they *really* what we've "needed and wanted"?

*What if* . . .

Instead of 1,400 more square feet, we'd been able to choose a modest increase, along with more eye-pleasing, comfortable, open construction, reducing heating and cooling costs—what's called "passive solar" that simply uses great insulation and takes advantage of the position of the sun to offer natural lighting and improve airflow and temperature?[25] Or an "active solar" home, using solar panels, which can sometimes generate more energy than it uses?[26] (Some good news is that in the aftermath of the housing collapse, the US trend toward ever-bigger homes is starting to reverse.)[27]

Or, instead of suburban homes with a lot more floor space, we could have opted for easy access to lovely parks and gardens where our kids could play outside safely? And fun points of neighborly connection, such as refigured libraries and community centers equipped with gyms, where we could hold club meetings, watch the World Series, or pump iron with friends, while our kids used the all-weather play spaces?

Or, instead of lives in suburbia, requiring long, frustrating commutes, we could have chosen close-knit, city living with rooftop gardens, where we could walk to entertainment, work, and parks, or get around quickly and safely on bikes or on modern buses enjoying dedicated lanes with weather-protected walk-on platforms, like those already taking off in Latin America?

And then there's the beef. What *might* we have chosen?

Several decades ago, as corn yields shot up, creating vast supply in a world with billions too poor to buy it, agribusinesses learned that keeping cattle in huge lots and feeding them tons of that cheap grain and soy (cheap in part because of almost $6 billion a year in subsidies) ended up fattening profits.[28] Advertisers then told us the resulting beef, "marbled" with more fat, was a tastier luxury than plain-old range-fed beef.

In 2002, cowboy poet Vess Quinlan reminisced about what happened after the sudden surplus of corn: "I grew up on the taste of range-fed beef, but soon I was hearing food celebrities patterned after Betty Crocker touting corn-fed beef on the radio, and I couldn't figure out what was wrong with beef raised on grass."[29]

The US Department of Agriculture did its part, too, to convince us. It labeled the fattiest beef the highest "quality" grade and called it "prime." Who wouldn't prefer that?

But, instead of swallowing these promotional pitches, what if we'd known that all the extra fat would worsen our disease risk, use vastly more grain and water to produce, and, for many eaters, be less tasty than range-fed beef anyway?

Political philosopher Benjamin Barber in *Strong Democracy* uses a cool metaphor that helps me grapple with Thought 2. In a market economy driven by the one rule of highest return to existing wealth (my shorthand, not Barber's), we "consumers" end up like diners in a giant cafeteria. We can point to the choices under glass and feel as if we have many appetizing options.

But what we *can't* do is get behind the counter, go into the kitchen, and plan the menus that include what we *really* want, including the discovery of some tantalizing suggestions from friends and neighbors that by ourselves we might never have dreamt up. As "menu planners" we could make choices based on what we gain and lose with differing options.

Perhaps including some of the "might-have-beens" above?

Plus, our sense of what we "need" turns out to be a lot less fixed than many assume: The share of Americans who feel air-conditioning is a necessity, for example, dropped from 70 percent in 2006 to 54 percent just three years later when times got tougher, and our attachment to dishwashers, clothes dryers, and microwaves weakened similarly.[30]

And our perception of enjoyment can make quick adjustments, too, I realized recently in this small example: Staying overnight at a friend's, I was handed a towel by my host. *How great*, I found myself thinking, as I felt a coarseness that I recognized immediately as the mark of line drying. *I love that roughness against my skin when I'm drying off.* And my next thought was, *Wait. When did I stop thinking downy-soft was best?*

Even long-held sensibilities about hygiene can transform rapidly. For Japanese, riding in a stranger's car has long been unappealing: "We like things clean and tend not to want to use things when we don't know who used them before us. So a car-sharing service will never be popular in Japan." That's what environmental activist Junko Edahiro got in response to her eco-appeal for car sharing a few years ago. But values are changing in Japan, and as a few tried car sharing, others got on board, and car sharing took off quickly. Now it's available at every station in Tokyo's main railway loop.[31]

And here in the US, more and more people are realizing there's a fun way to have what we need when need it, without rushing to the mall. Using the website tool NeighborGoods.net to find each other, people across the country had by 2011 shared goods—from a bread maker to a camping stove—worth $1 million. Clothing swaps are catching on, too. Recently, I loved watching my daughter's friends arrive at her apartment on a Saturday morning—each toting a bag of clothes they'd tired of—then with a lot of laughter slip into each other's offerings. Their affectionate name for their stuff-reducing fun? "Bitch and Grab."

In other words, recognizing that our needs and tastes are less fixed than we'd imagined and that so much is now known about how to align our desires with nature, we might choose very differently *if we had choice.* So the central question for me is not just, How do we buy less, but, Why don't we have choice, and how do we get it? Why don't we have the choice to create affordable communities aligned with environmental well-being in which we can experience ease, camaraderie, security, beauty, and stimulation—communities we can pass on with pride to our children?

And that question gets us digging deep, all the way to democracy itself.

## Waste before we buy . . . why?

We arrive at the question of democracy by considering perhaps the biggest reason we're in big trouble if environmentalism comes across as blaming consumer choices as the *root* cause of our crisis.

That reason is pretty obvious. If most of the waste and destruction occurs *before* we buy—if it is designed into the production process—then we could heed the call to cut our purchases way back—by, say, half—and still be massively disrupting natural regeneration as well as contributing to climate disruption.

What if there were just two hundred instead of over four hundred aquatic dead zones worldwide where sea life is being killed by farm-chemical runoff? Or what if the plastic soup smothering the Pacific Ocean were only half as big as it is—so it would be the size of only one Texas instead of two?[32] And what if there were only three instead of six pounds of plastic for every pound of ocean plankton in these heavily trashed areas?[33]

Half is still more than ecological rhythms can absorb.

Similarly, what if we could magically cut water-intensive food production back so we were depleting underground water in Midwest farm states at only half the present rate? The decline could still wipe out portions of this water source for farmers in fifty years.[34]

Neither would chastising us to buy less stuff, were it to work, touch other aspects of built-in destruction—such as that created by our military-dominated US economy in which the Department of Defense has become "the world's largest polluter, producing more hazardous waste per year than the five largest U.S. chemical companies combined."[35]

To get us on the path to a healthy planet, we have to get a handle on what drove us off the path. Taking us on a deadly detour have been many converging forces to which this brief chapter can only allude. But here's a start: The fear-of-lack-driven dogma of our "one-rule economy"—highest return to existing wealth—ends up relentlessly consolidating wealth, as stressed in Thought 1. And that concentrated private power has four profound consequences for "consumerism":

> *Enabling private influence over public choices.* With $3.5 billion spent in 2010 by mostly corporate lobbyists, private power shapes public policies, limiting choices consumers have.[36] The electric car, for example: Why did it take until late 2010 to see the all-electric Nissan Leaf in showrooms at $33,000 or the even pricier Chevrolet Volt? In fact, about fifteen years ago in California, General Motors *did* introduce an electric car, the EV1, to comply with that state's clean-air mandate. But, hit with lawsuits from carmakers, the oil industry, and the George W. Bush administration, California revoked the mandate—reveals the 2006 documentary *Who Killed the Electric Car?* In 1999, GM stopped production of its EV1.[37]
>
> How many of us are aware of this history? Or would connect the Pacific's plastic soup with unrelenting corporate resistance to rules requiring the recycling of plastic—despite the 30 billion plastic water bottles now tossed out, and mostly not recycled, each year in the US alone?[38]
>
> *Enabling ubiquitous and ever more sophisticated advertising.* On average, experts estimate that a typical American gets hit with 3,500 ads and promotions every day. Sound preposterous? Even if they're off by a factor of ten, imagine the power to endlessly shape how we perceive our choices.[39]

*Generating vast economic, educational, and medical inequalities.* Today inequality in the US is greater than in Egypt, Yemen, or Pakistan.[40] And we know inequality itself contributes to materialism, as those losing ground feel belittled and insecure and use purchases to gain "membership" and some easing of fear.[41]

*Denying Americans the fair, open, accountable governing process that's required to meet our challenges.* As I argue in Thought 7, no matter how much purchasing we resist, we can't meet today's crises and seize its opportunities unless we create a democracy in which our elected representatives answer to citizens.

## THE "GLORY OF THE HUMAN"?

What's right in Thought Trap 2 is that our culture's material preoccupation breeds unhappiness and ill health. What's not helpful is railing against consumerism as the cause of our crisis without exploring the real roots and solutions.

Trying to discourage shopping by intentionally frightening people with images of environmental collapse might trigger short-term behavior change, but, as in the psychology of dieting, a lot of people lurch from discipline to indulgence. Sustained change comes when we start to live differently from deep, positive emotions.

This morning I saw a big ad on the back of a city bus with this message: "Do something good for someone you love tonight." That "something good" turned out to be serving a vegetarian meal. The people behind this ad were telling us that serving loved ones food that's healthy, less hurtful, as well as less wasteful of the earth is a really positive thing to do. They were not playing on guilt; they were building on love. (To emphasize its benefits and to be more inclusive, I prefer to call my own vegetarian diet a "plant- and planet-centered" one.)

Similarly, creative efforts to halt companies from preying on young children are succeeding by refusing to guilt-trip parents.

Launched in early 2010, a campaign by Corporate Accountability International seeks to "Retire Ronald," the iconic clown of McDonald's, who's as recognizable to children around the world as Santa Claus. The campaign focuses not on blaming parents but—using petitions, shareholder actions, and more—on the unfairness and irresponsibility of advertising to susceptible kids products proven to create painful, lifelong health problems. One in three US children is predicted to develop diabetes in their lifetime.[42]

Corporate Accountability has a track record: In the 1990s, it targeted the Joe Camel icon because of its appeal to children—and won. The new campaign targeting Ronald McDonald is gaining traction.

When I asked my granddaughter Josie, at age three, how she thinks companies trick kids into eating stuff that's not good for them, without a second's hesitation she said, "Toys." Yep, she got it: McDonald's is among the world's largest toy distributors. (And Josie had never even been inside a McDonald's!) In 2010, the campaign against enticing kids like her to eat junk food succeeded in San Francisco when the city passed an ordinance restricting the use of toys or other "incentive items" to sell food that failed to meet clear nutrition guidelines.[43]

Surprised that the San Francisco supervisors had the gumption?

Consider this: In 2006, the city council of São Paulo, Brazil, the world's fourth largest metropolis, passed—forty-five to one—a complete ban on outdoor advertising. A year later 13,000 billboards were gone. Citizen support has been overwhelming, with 70 percent approving, and the council has resisted big advertisers' campaign to reverse the ban.[44] São Paulo isn't alone. In 2008, Buenos Aires advertisers and government agreed to get rid of 40,000 billboards, about 60 percent of the total.[45] And here at home, four states have banned billboards: Vermont, Maine, Hawaii, and Alaska.[46]

If my musings on the link between beauty and luxury reflect our real experience, then relief from visual clutter does more than give us a break from incessant solicitations to buy. Creating beauty in our urbanscapes— newly exposed sky, trees, and architecture—might help to satisfy our "luxury" desire without our buying a thing. And the billboard removal might do more than expose: In São Paulo, after the signs came down, some businesses painted their buildings more vivid colors.

So, let's resist naming our problem the "consumer society."

The framing itself suggests that our desires are in the driver's seat, or even on the throne—a note the late cultural historian Thomas Berry seemed to strike in 2006 when he said, "We might summarize our present human situation by the simple statement: In the 20th century, the glory of the human has become the desolation of the Earth."[47]

No, what we are living isn't the "glory of the human," as Berry of course understood. It's the degradation of the human—as we feel ourselves increasingly powerless and continue to inflict unnecessary stress and suffering on ourselves and other species. The real "glory of the human" can show up as we assert our power to create societies in which we can meet our desire for beauty and comfort in ways aligned with nature's gifts.

## THOUGHT LEAP 2

Let's not confuse symptom with cause. "Consumerism"—too much stuff!—is largely a symptom of forces denying us choice. Let's go for real choice as we delink what's become an automatic association between buying things, on the one hand, and luxury, comfort, and enjoyment, on the other. Together, let's imagine and create real luxury—rich, stimulating, and beautiful lives honoring the laws of nature.

thought trap 3:

# WE'VE HIT THE LIMITS
# OF A FINITE EARTH ]

*We've had it too good! We must "power down"*
*and learn to live within the limits of a finite planet.*

W E'VE BEEN LIVING BEYOND OUR MEANS FOR A LONG TIME AND now
it's all blown up in our faces," scolds Sir Jonathon Porritt, recent
head of Britain's Sustainable Development Commission.[1] Over the last
sixty years or so, we've all binged at a big fossil fuel party, we're told. Now
that party's over and, well, too bad, we must pull back.

Conveying the "power down" message quite vividly is Earth Hour. In
2008, from Sydney to San Francisco, people worldwide were encouraged
to turn out lights during the same hour. More than 50 million people
joined in. Since then, Earth Hour has gained enthusiasts, with people in
135 countries participating in 2011.[2]

Wonderfully, such massive involvement is yet more proof of people's
longing to be part of the solution. But does Earth Hour signal that the

climate crisis and the end of cheap oil mean more darkness, so let's all start getting used to it?

True, fear can motivate action, but it can also backfire. It's a "fundamental truth," says psychology professor Tim Kasser, that "when sustenance and survival are threatened, people search for material resources to help them feel safe and secure."[3]

As I argued in the previous chapter, insecurity can heighten fixation on material acquisition.

> I'd put my money on the sun and solar energy. . . . I hope we don't have to wait until oil and coal run out before we tackle that.
>
> —Thomas Edison, quoted in James Newton, *Uncommon Friends*, 1987

"We have a problem with Earth Hour," said student Victoria Miller at the University of Michigan, "because it suggests that the proper route to progress for humanity is shutting down and moving backward toward the Middle Ages." So Miller organized "Edison Hour," encouraging everyone to turn *on* lights to celebrate technology's contributions to progress.[4] The students' response suggests that, at least in a culture like ours, where we're encouraged to go it alone, "shutting down," as Miller calls it, can feel scary.

By the way, choosing the incandescent bulb, Thomas Edison's baby, as a symbol of progress is ironic, as it turns into light only 5 percent of the electricity it uses. Edison himself saw a lot of room for improvement.[5]

## THE LIMITS OF LIMITS THINKING

"Fossil fuels made the modern economy and all of its material accomplishments possible," writes the Worldwatch Institute, which I greatly admire, in its *State of the World 2008*.[6] And in their green economics textbook, *Ecological Economics*, environmental leaders and professors Herman Daly and Joshua Farley tell us that "fossil fuels freed us from the fixed flow of energy from the sun."[7]

Hearing these assessments, it is easy to assume, Whoa! Without fossil fuel, human ingenuity would never have come up with other ways to power our lives. The end of oil will mean giving up all the wonderful, modern "material accomplishments" that fossil fuel has made possible, as we get used to living constrained, once again, by the sun's "fixed flow."

But wait. Each day the sun provides the earth with a daily dose of energy 15,000 times greater than the energy humans currently use.[8] The sun is in fact the only energy that is not fixed in any practical sense. The energy of the sun is not even renewable. Rather, it is continually re-newing. We can't stop it!

But the biggest drawback of the "we've-hit-the-limits-of-a-finite-earth" idea is this: It frames the problem out there—in the fixed quantity that is earth. *Its* limits are the problem. This frame is carried, for example, in British environmental leader Tim Jackson's phrase "our ecologically constrained world."[9]

But, more accurately and usefully, the limit we've hit is that of *the disruption of nature we humans can cause without catastrophic consequences for life.*

The first frame conjures up the notion of quantity, as in a fixed but overdrawn bank account. The problem is the darn limit of the account, and the solution is to cut back what we withdraw. The second frame keeps attention focused on us—on human disruptions of the flows of energy in nature, which, if considered as systems, are renewing and evolving. Oil and coal, for example, are limited, certainly, but, as just noted, energy from the sun, for all practical purposes, is not. So, attention in this second frame is not on narrowly cutting back but on aligning with the laws of nature to sustain and enhance life.

## BEYOND LIMITS TO ALIGNMENT

If we conceive of our challenge as accepting the limits of a finite planet, our imagination remains locked inside an inherited, unecological world-view, one of separateness and lack. Precisely the thinking that got us into this mess. It's true, of course, that for all practical purposes our planet and atmosphere are made up of a limited number of atoms. But their configurations are essentially infinite. By conjuring up a fixed and static reality, the finite-limits frame draws us away from the deeper reality of our world—that of dynamism, which can offer stunning possibility if we learn to align with nature's rules.

*Think of music.*

Yes, there are just eighty-eight keys on the piano. But if we instruct ourselves to focus primarily on this limit, we won't get very far in creating beautiful sound. It is the possible variations on these eighty-eight keys that are important. And they are virtually endless; some are gloriously harmonious, others harshly discordant. Such quality is what must command our attention. A limits frame asks us to focus on the number of keys we use, but creating beautiful music requires deep learning of the principles of harmony. It requires both discipline and invention. Only by focusing on harmony can we know whether more or fewer keys are needed.

Making this core shift, we learn that, yes, we do uncover real limits on what we can do without disrupting nature's regenerative flows. But our sights remain clear: We make these discoveries as we focus on how our actions touch and are touched by all other life and as we continue to uncover and take inspiration from the laws of biology and physics.

We can learn, for example, how to cool our homes from a zebra's stripes.

Really. A zebra reduces its surface temperature by more than seventeen degrees Fahrenheit with microscopic air currents produced by the different heat absorption rates of its black and white stripes. In similar fashion, in Sendai, Japan, the Daiwa House office building uses alternating dark and light surfaces to create tiny air currents that control the building's exterior temperature. So indoor summer temperatures are lowered enough to save around 20 percent in energy use.[10]

## Waste not

Plus, once we see ourselves living within ever-evolving systems, our understanding of waste changes forever. We see that waste is not waste if it feeds an ecological process.

This holistic approach was dubbed "cradle to cradle" by William McDonough and Michael Braungart in their 2002 book by that name.[11] The term stuck. Cradle to cradle is the notion that, from buildings to upholstery to utilities, we can design productive processes so that their "waste products" feed other living processes rather than harm them.

And it's spreading fast, in part through efforts of the Geneva-based Zero Emissions Research Institute (ZERI) founded by green innovator Gunter Pauli. ZERI's motto: "Follow nature's example, realize waste's potential."[12]

A few years ago, in beautiful, mountain-ringed Manizales, Colombia, I got to see ZERI's vision coming to life when formerly jobless women showed me how they were earning a good income by using waste from local coffee processing as the substrate in which to grow highly nutritious mushrooms. The waste from the mushrooms then became feed for animals. This coffee-waste-to-mushrooms-to-feed connection has created 10,000 jobs in Colombia.

Imagine if the 16 million tons of waste now left rotting, and emitting greenhouse gases, on coffee farms around the world were instead feeding mushroom cultivation. Plus, if each of the roughly 25 million coffee farms in the world generated only two jobs growing mushrooms, says Pauli, coffee waste could provide 50 million protein-producing jobs globally. And tea farms could do the same with their waste, he adds.[13]

ZERI has spread this "pulp-to-protein" strategy to eight African countries. In Kenya, for example, water hyacinth—a vexing, foreign invader—has found an honorable calling as substrate that villagers now use to grow nutritious mushrooms, long part of local culture.[14]

To me, mushrooms have become almost magical in their powers: Scientist Paul Stamets, the mushroom magician, is showing the world that fungi can accomplish everything from killing termites to filtering toxins from farm waste to cleaning up oil spills—all by using nature's genius.[15]

Industrial ecology—one industry directly feeding another—is a step toward leaving behind the notion of waste. Another simple but powerful story comes from Japan, which in seven years cut municipal waste 40 percent: Professor Yoshihito Shirai of the Kyushu Institute of Technology became so distressed by the vast amount of food waste from the restaurant industry being carted off to landfills that he and his team of students and colleagues went to work. They came up with a way to use the discarded food—with help from a fungus (of course!)—to produce polylactic acid for bioplastic. It's done at nearly room temperature, saving energy, and the residue feeds animals.[16] Growing rapidly, bioplastics are mainly produced with fossil fuel–intensive corn, displacing food crops. Professor Shirai's approach makes a lot more sense.

## Garbage heat

Far from Japan, the 80,000 citizens of Kristianstad in a farming region of southern Sweden now use essentially no oil, natural gas, or coal at all to heat their homes and businesses—even through Sweden's long, cold winters. Two decades ago, fossil fuels supplied all their heat, but citizens of Kristianstad started to see farm waste—from potato peels to pig guts—with new eyes. Through a fermentation process, the city now generates methane gas, which then creates heat and electricity and even gets refined into car fuel.

"Once the city fathers got into the habit of harnessing power locally, they saw fuel everywhere," noted a *New York Times* account of Kristianstad's turnaround. So the city soon began taking advantage of waste wood from flooring factories and tree prunings to generate methane, as well as putting to use methane that was before being emitted into the atmosphere by an old landfill and sewage ponds.[17] (And because methane has even more potent greenhouse effects than does carbon, putting it to use is critical.)

"Waste to energy" is huge in Europe, with four hundred plants. Denmark is near the top, with twenty-nine.[18] By 2016 Denmark will be the "top" in a very different way. A futuristic, waste-to-energy plant in downtown Copenhagen, serving five municipalities, will generate heat and electricity for 140,000 homes, while doubling as a ski resort. In this flat city, skiers will be able to ride an elevator to the plant's "peak," then ski down its three encircling "slopes." Built into the design is a sobering lesson as well: The release of a visible smoke ring will be timed precisely so that onlookers can count five rings and know a ton of carbon dioxide has been released into the environment. The ring is intended as a startling way to make carbon dioxide real, motivating citizens to produce less waste to begin with.[19]

In the US, only a quarter of landfills capture methane from decaying garbage to make electricity, and even these emit over 50 percent more in carbon dioxide equivalents than waste-to-energy plants.[20] But imagine the positive potential: In the US, more than half of municipal solid waste—almost a couple of pounds for each of us each day—is just the kind of stuff used to heat Kristianstad. However, our waste is simply wasted, supplying

0.2 percent of our total energy demand.[21] Over half of our municipal waste goes to landfills—including 10,500 tons of residential waste leaving New York City every single day for landfills as far away as Ohio and South Carolina—a big contrast to Germany and the Netherlands, where roughly two-thirds of urban waste is recycled or composted, while only 1 to 2 percent goes into landfills.[22]

## Edison's other idea

Then there's wasted fuel itself.

Recall in Thought 1 that two-thirds of the potential energy in fuel that goes into a typical power plant is released as waste heat. Thomas Edison realized it didn't have to be this way and designed the world's first cogeneration plant on Pearl Street in lower Manhattan. That was 1882.[23] In cogeneration, a power plant's "waste" heat is captured and piped to heat or cool buildings or to power industry. Then, instead of two-thirds only 15 percent of the energy is typically wasted.[24] Notwithstanding Con Edison's thirty cogeneration plants now serving 100,000 Manhattan homes and buildings, this Edison invention hasn't taken off in the US . . . yet.

Its potential is huge. Cogeneration by itself could cut carbon emissions globally by 10 percent in twenty years, estimates the International Energy Agency. In Denmark it already provides over half of the electricity.[25]

These stories are a mere suggestion of the ways in which we're learning less about how to limit ourselves to stay within the earth's limits and more about how to harmonize our human systems with nature's ways.

As we become students of nature's laws, we find endless ways we can mimic the strategies of other creatures and plants to solve human challenges. In fact, what science writer and innovation consultant Janine Benyus has dubbed "biomimicry"—mimicking nature—is emerging as a new field of science. Engineers and architects are finding that even our most prized inventions are modest imitations of nature's feats: Lily pads and bamboo stalks mastered impressive structural supports long before human architects caught on. And the ability of termites to keep their towers at precisely eighty-six degrees Fahrenheit outperforms even our most powerful modern heating and cooling systems.[26]

## RIGHTING THE BALANCE

Another downside of narrowly focusing on reductions to stay within "limits" is that we're apt to miss a huge, crucial piece of the solution to the climate challenge.

Big mistake.

In the minds of most of us worried about climate change, averting catastrophe means cutting greenhouse gas emissions as fast as we can, mainly from their biggest current source—burning fossil fuel. That's essential. But, more accurately and usefully, we can frame our challenge as restoring a *balancing cycle* in nature.

"Carbon moves from the atmosphere to the land and back, and in this process it drives life on the planet," observes a 2009 Worldwatch Institute report. But we've been emitting much more carbon than our earth can reabsorb, throwing the cycle seriously out of whack. Our task now is restoring the "harmonious movement of carbon," the report concludes.[27] It's an example of what I mean by aligning with nature.

Greenhouse gas emissions now total roughly 47 billion metric tons of carbon dioxide equivalent* annually, and our earth has been absorbing about half, or about 25 billion metric tons.[28] Way out of balance! To close the 22 billion metric ton gap, re-establishing a balancing cycle of carbon as quickly as possible, we therefore need *both* to reduce emissions and to increase absorption of carbon each year.

Efficiencies, renewable-energy breakthroughs, and halting deforestation, along with shifts in our own perception of what makes us happy, highlighted in this book and many more, can reduce carbon emissions. *But how do we also enhance the equally critical carbon-absorption side of the cycle?*

To get a grip on why this question matters so much, consider what, for many, is a big surprise. It's possible that deforestation, farming, grazing, and other people-caused soil disturbance during prehistoric times put

* Carbon dioxide equivalent (CO2e) is an increasingly common measure of the global warming potential of greenhouse gases. It allows all such gases (including, for example, methane and nitrous oxide, which are much more potent per unit than carbon dioxide) to be counted and compared by a common standard: that of carbon dioxide.

more carbon into the atmosphere than has fossil fuel since 1850.[29] And even during the fossil fuel–intensive, post-1850 era, soil and plant disruption has released over one-third as much carbon as has fossil fuel.[30]

So, in righting the carbon balance, soil and plants have a big role to play. It requires both a "stop" and a "start":

We *stop* misusing rangeland and tearing down and burning forests. (The net loss of forests globally each year equals an area the size of Costa Rica, although the rate, still horrendous, has begun to slow.)[31]

And we *start* caring for soil, plants, and trees in ways that increase their carbon storing—some new ways, some *very* old. And some pretty simple: Lengthening the time between "harvesting" trees, for example, in "forests of the Pacific Northwest and Southeast could double their storage of carbon," notes the Union of Concerned Scientists.[32]

Better farming practices are just as central to our successfully rebalancing the carbon cycle. Today in the US, the food system contributes nearly a fifth of the country's greenhouse gas emissions.[33] Answers start with the dirt—no surprise once one learns that, overall, soil itself holds twice as much carbon as plants in the soil do.[34] Since both exposed and disturbed soils release carbon, the answer is farming in ways that avoid both as much as possible.

When using annual crops, that means not letting soil lie bare and instead planting cover crops, such as soil-enriching clover, in the gap between plantings of the annual crop. Better yet, it means relying more on perennials, including food-bearing trees as well as certain root crops and beans, so farmers don't have to disrupt the soil. Dr. Wes Jackson, the determined plant geneticist, and his team at the Land Institute in Kansas have strived for decades to develop perennial grains. They're getting closer, and their success could radically transform agriculture's negative eco-impacts.[35]

Climate-friendly farming also means forgoing chemical pesticides, as well as rotating crops and using compost, manure, and plants whose roots fix nitrogen, rather than applying manufactured fertilizers, to enhance fertility.

Agriculture contributing to a balanced carbon cycle also requires phasing out feedlots—now encouraged by tax subsidies—and moving livestock to well-managed range and pasture.

Until recently most worriers about carbon overload, including me, saw livestock as climate criminals, in fact among the worst offenders—now blamed for 9 percent of carbon emissions and 18 percent of all greenhouse gases measured in $CO_2$ equivalents. But here, too, some serious reframing is going on: It's not the animals that deserve all the blame, even though the livestock sector emits 37 percent of all methane, and methane packs a climate punch twenty-three times that of carbon dioxide.[36]

A big part of the problem is the way humans mismanage them: The largest share of carbon that livestock "cause" results from humans tearing down forests to create pasture and grow feed for them.[37] And add to that the climate costs of growing more than a third of the world's grain and about 90 percent of our soybeans—using vast amounts of fossil fuel—just to feed them.[38]

But livestock didn't ask to be penned up and stuffed with grain.

Proof is trampling in from Australia and Africa that carefully managed grazing animals can help the earth absorb carbon. Despite widespread overgrazing, speeding desertification and releasing carbon worldwide, livestock could actually help reverse the process: They can break up hard-packed earth, deposit manure, enable seeds to take hold and water to penetrate, and, without even trying, regenerate healthier grassland and waterways—absorbing significant amounts of carbon.

But for that to happen, humans would have to learn to herd the way nature used to: From time immemorial, natural predators have forced animals into groups and kept them moving often, and now herdsmen are learning to mimic the approach. They bunch animals together and leave them no longer than three days on one piece of land.

While school kids now know that forest vegetation stores carbon, it turns out that the grassland stores as much, mainly in the soil, so the potential impact of this breakthrough—what renowned innovator Allan Savory calls "holistic, planned grazing"—is big.[39] Worldwide, grazing land covers more than a quarter of all ice-free terrain, 8 billion acres or more. But so far this low-cost, holistic, carbon-absorbing path to grassland restoration has only reached 30 million.[40]

Imagine the possibilities if we shifted public support to such efforts: Even without counting what this grazing breakthrough could mean, experts report that these very doable farming practices cooperating with nature to grow our food—called agroecology—have the "technical potential" to absorb up to 6 billion metric tons of carbon dioxide equivalent each year by 2030, *or roughly a quarter of what's needed to achieve carbon balance.* And some experts say the potential is much greater.[41]

We certainly don't want to miss that.

One reason agriculture can become such a big piece of the climate-stabilizing puzzle is that growing trees and shrubs among food crops is not a problem. It's a really good thing. Called "agroforestry," the practice can improve productivity not only because the trees help keep soil from being washed or blown away but because the roots help water penetrate the soil. Plus, some tree varieties "fix" atmospheric nitrogen in the soil, effectively producing their own fertilizer. Farms with these "fertilizer trees" mixed in among field crops double or triple crop yields, reports the World Agroforestry Centre, while at the same time cutting the use of climate-disrupting commercial nitrogen fertilizer by up to 75 percent.[42]

## "WE STOPPED THE DESERT"

Consider the impact in West Africa, where in many minds climate change and deep poverty meld into heartbreaking images of destitution on increasingly scorched earth. Indeed, three-fourths of Niger is now desert, and the only news we heard from the country in mid-2010 is that famine threatened half of its people.[43]

Grim . . . yes?

But there's another story. Over two decades, poor farmers in the country's south have "regreened" 12.5 million desolate acres, a momentous achievement not of planting trees but abetting their "natural regeneration." There, a farmer-managed strategy has revived a centuries-old practice of leaving selected tree stumps in fields and protecting their strongest stems as they grow. The renewed trees then help protect the soil, bringing big increases in crop yields, and they provide fruit, nutritious leaves, fodder, and firewood.

In all, Niger farmers have nurtured the growth of some *200 million* trees.

In the mid-1980s, it looked to some as though Niger would be "blown from the map," writes Chris Reij, a Dutch specialist in sustainable land management, but farmer regreening has since brought enhanced food security for 2.5 million people.[44]

So, in late 2010, even as many in Niger were facing shortages, village chief Moussa Sambo described his village near the capital as experiencing the greatest prosperity ever, with young men returning. "We stopped the desert," he said, "and everything changed."[45]

And why hadn't hungry farmers in Niger figured this all out long ago?

Well, they had. But in the early twentieth century, French colonial rulers turned trees into state property and punished anyone messing with them.[46] So farmers began to see trees as a risk to be avoided and just got rid of them. But Niger gained its independence in 1960, and over time, Reij says, farmers' perceptions changed. They feel now they own the trees in their fields.[47]

And why haven't we all heard about their extraordinary achievement?

The whole of southern Niger "was assumed to be highly degraded. . . . Few thought to look for positive changes at a regional scale," Reij notes.[48] And "if people don't know to look for it, they don't see it."[49]

Could this be yet more evidence of our mental map's filter working against us?

Now aware, though, we can take heart from African farmers' creativity in the face of a deteriorating environment, and they're hardly alone.[50] If proven agroforestry practices, like those in Niger, were used on the over 2 billion acres worldwide where they're suitable, in thirty years agroforestry could have a striking impact—accounting for perhaps a third of agriculture's overall potential contribution to righting the carbon balance.[51]

Beyond agriculture is the larger potential of forests.

Reducing our current forest destruction, planting new forests, and improving how we manage forests could sequester almost 14 billion metric tons of carbon dioxide equivalent a year by 2030, says the Intergovernmental Panel on Climate Change.[52] So, adding this 14 billion potential contribution of forests to agriculture's 6 billion potential, we're approaching the brass ring: closing that 22-billion-metric-ton gap between the carbon dioxide equivalents we're now emitting and what earth must absorb to avoid catastrophe.

## From ancient farmers, a soil secret

Another dramatic climate-helping, soil-enhancing breakthrough is nothing new at all: It's an ancient Amazonian practice of smoldering organic waste to create a form of charcoal that's added to the topsoil.

Now called "biochar," its secret is its porous structure, which is welcoming to the bacteria and fungi that help plants absorb soil nutrients. So, biochar added to soils typically increases crop yields, sometimes even doubling them.[53] And it is great for poor farmers because it can be made from material that otherwise would be discarded—in Africa, for example, cassava stems, oil palm branches, and common weeds. The controlled smoldering required to make biochar can also generate clean energy, obviating the need to cut down the forest for firewood. Plus, producing biochar removes carbon from the atmosphere and can lock it away for centuries. Biochar's promise is being explored in test fields from Iowa State University to villages in the Congo.[54]

It's a breakthrough worth following with an eco-mind that knows context is crucial: Even biochar could harm those less powerful, if agribusiness is allowed to create huge biochar operations displacing them.[55]

An eco-mind sees that balancing the carbon cycle, while enhancing fertility and yields, is largely about spreading proven *practices* available to almost all farmers, not new *purchases* available only to a minority. It focuses on empowering relationships—resisting technologies, including genetically modified and other patented seeds, that make farmers dependent on distant suppliers.

What's great is that balancing the carbon cycle and helping the poorest farmers calls for the same public actions: We shift support from fossil-fuel intensive farming toward agroecological practices. We take strong action against deforestation while supporting massive tree-planting initiatives, as in Ethiopia, and fostering trees' "natural regeneration," as in Niger. With an eco-mind, these steps—both cutting carbon and storing more—are urgent and satisfying.

## HUNGER AS TEACHER OF THE ECO-MIND

The danger within the "limits frame" first hit me when I began asking, How do we end hunger? I realized that humanity has long seen the solution as getting the *quantities* right—making sure the quantity of food can feed the "quantity" of people. And we've done it.

We've succeeded in both growing more food *and* slowing population growth. But, still, 868 million people go hungry. And this "official" count needs a hard look. To be counted "hungry," a person has to survive for more than a year on less than the minimum calories required for a "sedentary lifestyle." I was shocked. Poor people in developing countries are likely among the world's *least* sedentary.

So what if the UN hungry-people counters had instead used their definition of "normal activity"? Hungry people would almost double, to 1.5 billion.[56]

And because we humans tend to see what we expect to see, it's easy for us to see so much hunger and blame "too little food and too many people," whether true or not. In the summer of 2009, a *National Geographic*'s cover story "The End of Plenty" stated flatly: "For most of the past decade, the world has been consuming more food than it has been producing."[57] Even the brilliant environmental leader Bill McKibben suggests that climate change is already denying us the quantity of food needed.[58]

So of course we'd assume humanity has overrun Earth's finite capacity and our only hope is fewer people. But we'd be wrong.

Yes, of course, our birth rates must come into harmony with the earth, and that can happen as we tackle the root cause of population growth—the same power imbalances in human relationships that create hunger. Note that 95 percent of population growth is in poor countries, where the majority, especially women, lack sufficient power over their lives.[59]

But a "not enough" diagnosis ignores this even more obvious fact: Even though the world's population has nearly doubled since the late 1960s, today there's significantly more food for each of us, reports the UN's agricultural arm: now almost 3,000 calories per day. That's plenty—and, remember, it's only with the leftovers: what's left over after we feed more than a third of our grain and most of our soy to livestock. Over the last decade, even the fifty "least-developed countries" as a group have experienced per-person food production gains.[60]

So *National Geographic*'s scary declaration belies the facts. *Hunger isn't the result of a lack of food.* And thus a simple frame of "hitting the limits" can't help us understand what's going on.

We need an eco-mind that never stops asking why.

"Since the early 1990s, food[-import] bills of the developing countries have increased by five-or six-fold," notes Olivier De Schutter. And he should know, for De Schutter is the UN Special Rapporteur on the "right to food." He emphasizes, though, that this deepening dependency reflects powerful human-made forces, including foreign aid and local governments' defunding agricultural development, including agriculture extension agents.[61] One reason is that foreign aid to poor countries was often tied to their governments' opening doors to imported food and cutting public supports. Sound familiar?

So agriculture in many poor countries faltered, and millions of farmers abandoned the land for urban centers. Cities grew, and poor city folk couldn't find decent work, so their lives depended on cheap food. Feeling that pressure, governments have tried to keep food in cities cheap, which depends on further undercutting profits farmers need to invest in producing more. Desperate governments opening their doors to cheaper imported food only made it harder for their own farming to flourish. Speeding the cycle, governments in the Global North didn't follow their own advice, and continued to subsidize their farmers big-time. So their artificially cheap grain exports also encouraged import dependence in poor countries. At the same time, corporate control over seeds and farming supplies has been tightening, leaving farmers with a shrinking share of the return from farming.[62]

And, as if these extreme power imbalances weren't bad enough, there's Wall Street's entry.

Over just three years, from 2005 to 2008, the price of hard red wheat, to pick one example, jumped fivefold—even though wheat was plentiful. What had happened? In 1991, Goldman Sachs, followed by other banks, Started putting investor money into their new commodity indexes—where dollars invested have ballooned fifty-fold since 2000, explains Frederick Kaufman in *Foreign Policy*. In what he calls a "casino of food derivatives," speculative dollars overwhelmed actual supply, and in just three years, 2005 to 2008, "the worldwide price of food rose 80 percent."[63]

And it's only gotten worse.

During much of the last two years, the UN Food Price Index has been roughly twice as high as a decade ago, unleashing a long-term, hunger-making force: In an era of rising food prices, speculators and governments worried about their populations' future food supply—including the Gulf States, South Korea and China—are seizing cheap land.

In 2009, land purchased by speculators and foreign governments, especially in Africa, jumped more than tenfold (to about the size of France) compared to previous years, reports the World Bank. They're buying especially where governance is "weak," the Bank notes; thus making it easier to get land "essentially for free and in neglect of local rights."[64]

Imagine our feelings of vulnerability if this loss of control were happening to us.

Other factors have played, and continue to play, a role in both food-price escalation and price swings, including worsening climate-change-related flood and drought, the rising price of oil, world food reserves allowed to sink too low, along with government-mandated diversion of grain into making fuel—*in the US enough in sheer calories to feed a population larger than ours.*[65]

Thus, the continuing tragedy of hunger, during an extended period of largely excellent world harvests, stems overwhelmingly from concentrated economic power.[66]

My point is that fixation on quantities and limits makes us eco-blind, unable to see, and therefore not driven to explore, key human relationships—in this case, from those setting off food-price escalation to those enabling people to choose the size of their families. All make up our social ecology, determining who has the power to eat. The mechanical, quantitative view keeps us from seeing that in both human and nonhuman realms, relationships have become so mal-aligned, so unharmonious, as to generate vast hunger—even amid unprecedented food abundance.[67] So, the useful questions are about the re-alignment of our most basic relationships. They are:

*Do our methods of production enhance ecological relation-*
*ships that restore and maintain food-producing capacity as they*
*help to rebalance the carbon cycle?*
*    And do our human relationships enable all people to gain*
*access to what is produced?*

Diverted from these questions by thinking within a simple, mechanical frame of "more or less," we can't see that the very strategies we've used to grow more have ended up so concentrating power over food that hundreds of millions go without. The frame has kept us blind to an entirely different approach already flourishing in diverse settings—an approach focusing on dispersion of social power as we cooperate with nature, one through which all of us can eat well while enhancing soil and water quality.

Think back, for example, to the farmers' breakthroughs in Andhra Pradesh, India, or in Niger. Not by focusing narrowly on "more" but by radically and positively remaking their relationships to the land and each other, they're gaining ground both in meeting food needs and in creating healthier communities.

## FLOURISHING AS, OR EVEN BECAUSE, WE CUT GREENHOUSE GAS EMISSIONS

Given all we now know, why, I often ponder, aren't we in the midst of exciting national discussion about how quickly we can leave fossil fuel behind?

One obstacle might be an unspoken notion that if we're not doing something we "should be," the reason has to be that it costs too much. Since we're not responding to the threat of climate chaos, it must be that the price tag is too high. So we can't see that what's hugely expensive is inaction, whereas action will save us vast sums. Or maybe our country's Puritan heritage is still whispering to us that doing what's right has got to hurt. And we don't want to hurt; we're already hurting too much.

This "the-party's-over" thought trap might reinforce these perhaps less-than-conscious assumptions, blocking us from realizing that cutting greenhouse gases can enrich many aspects of our lives.

Here are just some of the ways:

*We'd certainly save money.*

The Union of Concerned Scientists "blueprint" shows how in two decades, primarily via renewable energy and advances in efficiency, we could cut carbon significantly and at the same time end up saving the average US household $900 on electricity and transportation a year. By 2030, overall, Americans would experience a net gain of $464 billion annually.[68]

Buildings offer huge potential for energy savings, since they account for more than a third of US energy use. Consider the Empire State Building, where investing in efficiencies is projected to reduce by 40 percent its $11 million yearly energy outlay, reports Amory Lovins's Rocky Mountain Institute. Strategies include superwindows six times more efficient than regular double-paned windows and insulated barriers placed behind radiators to reflect heat.[69]

In similar redesigns across a wide range of industries, Lovins's team consistently finds energy savings of 30 to 60 percent in old plants, paying back the investment in two to three years, and 40 to 90 percent in new plants.[70] A sixth grader could grasp some of the money-saving energy-efficiency schemes. Lovins notes, for example, that 60 percent of the world's electricity runs motors, and the biggest use of motors is for pumping. Out of pumps come pipes, and Lovins finds that cheaper, low-friction pipes can save as much as 92 percent of the pump's energy.

The trick? Replace "skinny, long, crooked pipes" with "fat, short, straight pipes. . . . This is not rocket science," says Lovins.[71]

Such is a taste of the kinds of savings within reach. And if one still doubts the big efficiency gains available to us, take note: Other countries are already far down the road. Ireland and Switzerland generate twice as much production as we do for every unit of energy used.[72]

*And meeting the challenge of up-front investment required?*

In 2008, the research arm of eighty-two-year-old management consulting firm McKinsey & Company found that, globally, "the costs of transitioning to a low-carbon economy are not [economically] all that daunting." The study estimates that the US could fund a low-carbon economy mostly "from investments that would otherwise have been made in traditional capital."[73] Globally, investing $170 billion each year in energy efficiency would bring an "energy savings ramping up to $900 billion annually by 2020," concludes another McKinsey report.

And investors would get a 17 percent rate of return.[74] Not bad.

*And jobs?*

Moving toward electricity from wind, solar, and biomass could provide three times the number of jobs compared to continuing dependence on coal and gas, finds the National Council for Science and the Environment.[75]

*And health?*

One measure of the vast health dividend we can enjoy as we move away from fossil fuel is captured in part within estimates of the hidden costs of coal, reported in the major new study cited earlier. In illness, lost productivity, and more, these costs come to $269 billion each year.[76] Imagine being free of that burden.

Food offers another enticement to embracing the sun's energy. Here the alignment between what's good for our bodies and what's good for the earth—plus other creatures on it—is stunning. My daughter, Anna Lappé, brings to life in her 2010 *Diet for a Hot Planet* how earth-friendly, family-scale farming captures all the "efficiencies of scale" while creating healthy soil, water, more and better jobs, and healthier food. Not only does eating food produced organically, especially fresh and whole food, encourage modes of production that reduce climate impacts, but we eaters avoid toxic chemicals and highly processed products—saving ourselves from a diet that's become a major health hazard (with costs rivaling that of tobacco-related disease).[77] Plus, we get on average a quarter more nutrients per bite than if eating produce grown using farm chemicals.[78]

Now there's a win-win.

And, to help us see these gains, Hollywood is pitching in too: "You don't even have to believe in the existence of climate change to understand that an energy revolution may be the very thing we need," says TV and movie producer Marshall Herskovitz, who's leading an entertainment industry initiative to open Americans' eyes to the benefits of moving beyond fossil fuel. "We are in a very rare moment in history where the solving of one problem would actually solve four or five or six other intractable societal problems we have in the United States—unemployment, the deficit, our trade deficit, health, national security."[79]

## Have fossil fuels freed or enslaved us?

Yet, within the limits frame, the opposite seems to be assumed—that fossil fuel temporarily removed constraints so we could indulge ourselves. We're told that we are "addicted to oil," as if on a drug high from which we now must descend. In fact, many people promoting a post–fossil fuel world use the term carbon "descent" to capture what's now required of us.

So, here's the snag: When economists write that "fossil fuel freed us," they make it easy to forget that fossil fuel has also entrapped us. Because it exists in concentrations, fossil fuel has inexorably fed the concentration of social power in the hands of the few with the resources to extract it and to make the rest of us their dependent customers. That power means profits. Exxon's almost doubled in just four years, to more than $45 billion in 2008, even as much of the world was devastated by the financial crisis. That's $1,434 a second![80]

Such highly concentrated power, as we've long known, typically leads to really bad things—cruelty and suffering among them. Consider Nigeria. "Everything looked possible" for Nigeria, writes Tom O'Neill in *National Geographic*. Then oil was discovered in 1956, and "everything went wrong," as he captures in these scenes of Nigeria today:

> Dense, garbage-heaped slums stretch for miles. Choking black smoke from an open-air slaughterhouse rolls over housetops. Streets are cratered with potholes and ruts. Vicious gangs roam school grounds. Peddlers and beggars rush up to vehicles stalled in gas lines. This is Port Harcourt, Nigeria's oil hub. . . . Beyond the city . . . exists a netherworld. . . . Groups of hungry, half-naked children and sullen, idle adults wander dirt paths. There is no electricity, no clean water, no medicine, no schools. Fishing nets hang dry; dugout canoes sit unused on muddy banks. Decades of oil spills [by one estimate, equal to an *Exxon Valdez* spill each year for over fifty years],[81] acid rain from gas flares, and the stripping away of mangroves for pipelines have killed off fish.[82]

Nigeria is the world's seventh largest oil exporter, earning the country nearly $60 billion a year, yet it so lacks refining capacity that it must import fuel, and its annual per capita income is less than that of nearby Senegal, which exports not oil but fish and nuts.[83] Nigeria's poverty is so great that life expectancy there, forty-seven years, is among the world's worst.

Oil wealth breeds a deadly antidemocratic unity of foreign corporate power interested only in protecting its profits and local government corrupted by the huge sums it can pocket by cooperating with the oil companies.

Royal Dutch Shell, for example, has dominated oil extraction in Nigeria since the late 1950s. Recently, the company agreed to settle out of court a lawsuit by victims' families and the New York–based Center for Constitutional Rights, which accused Shell of colluding with the Nigerian government to abuse human rights. Denying any guilt, the company paid out $15.5 million—or about four hours' worth of its 2008 profits.[84]

In countries where oil is concentrated, "freedom" and "oil" operate in "an inverse correlation," notes *New York Times* columnist Thomas Friedman.[85]

## AND HOW ELSE HAS OIL ENSLAVED?

Here at home, whether or not you believe that the drive to control oil lies at the heart of the $1 to $3 trillion US-initiated war in Iraq, it is unarguable that a fear of losing control of oil drives key aspects of US foreign policy. How could it not? The thirteen-member Organization of Petroleum Exporting Countries—half of which are in the Middle East—controls about half of the world's oil, and we depend on this cartel for 40 percent of our crude oil.[86] How can any nation feel free and confidently plan for its wellbeing if dependent on imports for essential energy?

Concentrated social power—flowing inexorably from the physical concentration of fossil fuel and the concentrated wealth it takes to extract it—undercuts democracy in yet another way: As long as we allow private wealth to influence campaign outcomes and infuse itself into public policy making, Big Oil will continue to throw its gargantuan resources behind policies favoring it at the expense of the planet. Just one galling example: Despite our climate crisis, $300 billion in annual global energy subsidies continue mostly to promote planet-heating fuels.[87]

For years, US oil and gas companies have wrangled major exemptions from laws, including the key Clean Water Act, that might have protected our water from the toxins they use in drilling.[88] Perhaps with BP's recklessness—abetted by lax government oversight—now exposed in the tragic 2010 Gulf of Mexico oil gusher, more Americans will awaken to the downside of oil dependency—*if* we can make clear that a safer alternative path is truly viable.

The concentrated power flowing from fossil fuel also gives those who control it so much wealth that they have plenty to put toward confusing us—for example, by purchasing $50,000 ads in the *New York Times* on the opinion page, which readers associate with ideas, not advertising.[89] There, in June of 2009, for example, ExxonMobil bragged that it had invested $1.5 billion over the previous five years to decrease emissions and increase energy efficiency. What readers weren't told was that in 2008 alone, the company spent $26 billion—seventeen times more—on oil and gas development.[90]

And Exxon's research on renewable energy? In 2008, Exxon spent $4 million (that's an *m*, not a *b*) on renewable-energy research.[91]

From their claims, we'd never guess that during the last fifteen years the top five oil giants, with roughly $80 *billion* in combined profits in 2008 alone, provided only about a tenth as much capital for clean energy as have venture capitalists and other corporate investors.[92] At the same time, they've helped to confuse citizens about climate change and spread the "government-is-our-problem" philosophy to disempower our democracy.

The oil giants are in the way of, not part of the way toward, life.

Finally, since security is foundational to democracy, fossil fuel dependency undermines democracy in yet another way. Former director of the Central Intelligence Agency James Woolsey nailed it when he noted that in the US "our focus on utility scale power plants instead of distributed generation" makes our energy grid "vulnerable to cyber and physical attacks." He called on us to boost distributed power generation from wind and solar.[93]

Considering all this, might our descendants look back at this era of The End of Oil and conclude that it marked the beginning of real freedom? With hindsight, will they see that as humanity moved to rely on the sun's distributed energy, social power became more distributed too—and that this shift was a necessary antecedent of real democracy?

## DISTRIBUTING SOCIAL POWER AS
### WE GENERATE NEW AND CLEAN ENERGY

Unlike fossil fuel, solar energy in all its forms gives most humans the chance to be cogenerators. For the biggest "waste" in today's world is that of the sun's rays. Less than five days of the sun's energy is greater than all proven reserves of oil, coal, and natural gas.[94]

Consider Denmark. Its early experience with wind energy—a form of solar power itself, since wind results from the sun heating the air—offers a taste of how humans can use the sun's distributed energy and keep social power distributed as well.

In 1980, Denmark introduced a 30 percent subsidy for investing in wind power. Partly as a result, cooperatives, made up of a few individuals or a whole village, helped turn Denmark into a world leader in wind energy. Cooperatives now own about a fifth of Danish wind power. Denmark's policies ended up encouraging 175,000 households to become producers, not just consumers, of energy—either through individual or cooperative ownership.

This direct citizen involvement changed Danes' perceptions.

With a stake in the wind installations themselves, producer families accepted their altered landscapes. But when government support for distributed production waned and "larger, purely business investments" came in, the "public became less willing to look at wind turbines."[95] The shift in perception highlights a common human experience: that what we ourselves choose and create we see through different eyes than if the very same thing had been imposed on us. This insight seems key to transforming resistance in the US, where big wind projects, most notoriously Massachusetts's offshore Cape Wind, have met mighty opposition.

And how has Denmark become a world leader in renewable energy?

Jane Kruse says it started with regular citizens. Jane directs a center for renewable energy in one of her country's poorest areas and credits "young people and women [who] were very vocal against nuclear energy."

Momentum grew steadily through the 1970s and early 1980s, she says, until in 1985 the Danish parliament decided to build no more nuclear reactors. In an interview at Wind-Works.org, Jane adds, "But, we were not only struggling against nuclear, we also wanted to work for positive alternatives." So women politicians (now more than a third of the parliament) joined to oppose nuclear energy and "cooperated across parties to pass legislation supportive of renewable energy."[96]

In Germany, too, everyday citizens stepped up. In the Black Forest community of Schönau, Ursula Sladek, a mother of five, was shaken up by the 1986 Chernobyl nuclear accident. She, like Jane, decided not just to fight nuclear power but to create an alternative. By 1997, she and neighbors had raised the millions of euros needed to buy out the area's private power grid and turn it into a co-op. Now owned by more than 1,000 people, it uses and supports decentralized renewable power, including solar and wind, to serve 100,000 customers, including both households and factories. It all got started because one woman said "no"—*and* "yes." Now all Germany is with Ursula, rejecting nuclear power."[97]

In the early 1990s, Germany had virtually no renewable energy, but now the country gets 16 percent of its electricity from renewables and is on track to achieve 35 percent within ten years.[98] Germany's policy, now spreading worldwide, is called the Feed-In Tariff because producers receive a payment ("tariff") for feeding clean energy into the energy grid. The law obligates utilities to buy electricity from renewable installations, like a solar panel or small windmill, at a price that guarantees a good return.

German households seized the opportunity and now own roughly 80 percent of the country's solar installations as well as most of its small hydroelectric power plants. The cost of the whole program is spread across all ratepayers, coming to less than $5 a month per household—all while stimulating 370,000 jobs in the renewables industry.[99]

This practical scheme for distributed power generation is now working in dozens of countries on six continents.[100]

Yes, experts tell us, to fully embrace the dispersed sun, wind, and other clean-energy possibilities, we'll also need to invest in what's called a "supergrid," connecting and balancing demand through dispersed green power generators. If we let it happen, concentrated social power—

those companies wealthy enough to invest in grids—could gain ground in a new form. But it's not a given. As more of us become energy generators ourselves—picking up the spirit of Jane and Ursula, in ways impossible with fossil fuel—isn't it likely that we'd resist a return to dependency?

## A DIFFERENT PATHWAY, A DIFFERENT MESSAGE

Of course, only a portion of the vast potential suggested here, in everything from "natural regeneration" of trees, to biochar enhancing of the soil, to impressive energy efficiencies and distributed energy generation, is practically achievable any time soon. But their potential is so far beyond what's required that a "portion" would be terrific.

My concern, however, is that a frame of "limits" can limit our view—keeping us from seeing the many positive steps we can take right now to balance the carbon cycle.

The 2009 Union of Concerned Scientists peer-reviewed study *Climate 2030: A National Blueprint for a Clean Energy Economy* would put us on the path to cut climate-disrupting emissions by 2050 to 80 percent below their level in 2005. Is it enough?

The Copenhagen Accord, signed by 167 countries, says that to avoid catastrophe we must keep planet-heating below 2 degrees Celsius (3.6 degree Fahrenheit). But even if we stopped carbon emissions now, reports climate-change fighter Bill McKibben, our prior actions mean we can't avoid a planetary temperature rise approaching 2 degrees. Worse, burning remaining fossil fuel could release carbon propelling us five times beyond the 2 degrees. It's "terrifying math," says McKibben.[101] And it is.

Our response can be to freeze in fear or to use this new knowledge to motivate us to implement with even-greater vigor the many known strategies for reducing emissions and holding more carbon in the soil and plant life.

To do so, though, we need very different messages. Just as in Thought 2's "the-consumer-is-to-blame" refrain, a "the-party's-over" framing of our challenge is a big nonstarter for many. In 2008, British prime minister Gordon Brown dubbed what we've been living the "age of global prosperity." Oh yeah? Most people didn't feel they'd been invited to that party, even before the Great Recession. The financial stress many Americans feel well predates the most recent crisis: The bottom 90 percent of us, as noted earlier, were already earning less in real dollars than in 1973.[102]

We defeat our ends if environmental messages make already-stretched families fear that protecting the environment means losing further ground. An understandable response might be to grab everything in sight, *now*, before it's all gone.

So, let's strive for a vision of less pressure and more security.

"The place we could finish up could be so much nicer than the one we've got now," says Tony Juniper, once director of Friends of the Earth, UK, and now a leader in an international movement called "Transition Towns." "We're not headed back to a new Stone Age or Dark Age, we're headed toward a much brighter, secure future, where communities are re-built, pollution is a thing of the past, we've got food security, biodiversity, people have long comfortable lives, energy is secure forever."[103]

No doubt this spirit is a key to why the Transition Towns initiative is taking off. It was launched only six years ago in Kinsale, Ireland, by eco-farming and gardening educator Rob Hopkins. Rather than as threatening a scary time ahead, Hopkins sees the climate challenge as an "extraordinary opportunity to reinvent, rethink and rebuild."[104] It's an "experiment in engaged optimism," he says.[105]

The movement has become a network of communities pledging and plotting to transition to renewable energy, while re-creating local economies and other aspects of community well-being. In addition to the almost four hundred "official" Transition Towns already participating in fourteen countries, many hundreds of other communities have expressed strong interest. And thousands of communities see themselves as part of the movement, says its founder. A couple of Transition Towns in the UK have even created their own green energy utility companies, and the Scottish government is helping fund local Transition Movement initiatives as part of its official response to climate change.[106]

The Transition Towns movement's slogan of "carbon descent" might more appropriately be "carbon freedom," for Hopkins's message and the movement's spirit capture a way of seeing that ignites human imagination and invention. Who wouldn't want to be part of his "experiment in engaged optimism"?

## THOUGHT LEAP 3

Because most people know they weren't invited to the Too Good party, the message of limits falls flat. An effective and ecologically attuned goal is not about more or less. Moving from fixation on quantities, our focus shifts to what brings health, ease, joy, creativity—more life. These qualities arise as we align with the rules of nature so that our real needs are met as the planet flourishes.

Aligning with the laws of nature means re-embedding our societies, including economic life, in democratic relationships of mutuality. Yet, here we come up against a fourth thought trap, a dim view of human nature itself.

thought trap 4:

# WE MUST OVERCOME
# HUMAN NATURE
# TO SAVE THE PLANET

*Humans are greedy, selfish, competitive materialists. We have
to overcome these aspects of ourselves if we hope to survive.*

"WE EVOLVED TO SUIT A WAY OF LIFE WHICH IS ACQUISITIVE, territorial, and combative," proclaims popular British writer Philip Pullman, and now we have to overcome "millions of years of evolution."[1] Renowned biologist E. O. Wilson warns us that our "Stone Age emotions" are an obstacle to saving our "living environment."[2]

Wow. If I believed all of this, I would feel utterly hopeless. How can we possibly make the needed planetwide turn to align with nature if first we have to change our own nature?

Here appears the second huge piece of the "premise of lack" dragging us downward. First was the lack of goods—from food to energy—and now we meet its partner, our lack of goodness. To believe that we are nothing but selfish little shoppers, we have to deny our own everyday experience. Even a moment's reflection—and now a lot of science—suggests that humans are much more complex creatures.

So, we can drop the debate over the inherent goodness *of* human nature—agreeing that humans can be both selflessly giving and cravenly cruel. We are then able to identify the deep positive needs and capacities *in* human nature that we can tap for all to thrive.

And it matters. It really matters. How we think about who we are has tremendous power over how we act. And some clever scientists with a bunch of Chloé sunglasses recently revealed how shockingly true this is.

First, they handed out the pricey sunglasses to a group of women, telling half of them at random that they were really only getting fakes. All were then assigned identical math problems for which correct answers brought money rewards and asked to score themselves. The researchers found the rate of cheating differed big-time: 70 percent of those believing they were wearing fakes cheated—a rate more than double that of subjects who believed their glasses were "real." Apparently, merely perceiving themselves as inauthentic led the bigger cheaters to act more unethically.[3]

If how we view our own nature shows up dramatically even in a trivial experiment, isn't it high time we carefully examined our inherited, damning views of our *whole species*?

From works like *The Naked Ape* in the 1960s to *The Selfish Gene* in the 1970s—not to mention the nightly news—a lot of us have learned that our species is inherently and basically brutal. But that view defies the evidence: "There is a very tiny handful of incidences of conflict and possible warfare before 10,000 years ago," says archaeologist Jonathan Haas of the Field Museum in Chicago, "and those are very much the exception."[4]

Our oldest known human ancestor, *Ardipithecus ramidus*, or "Ardi," who lived 4.4 million years ago, was only discovered in 1994 in Ethiopia. She's produced a "tectonic shift" in views of human evolution, says anthropologist Owen Lovejoy of Kent State University. "We now know . . . that hominids have always been a far less aggressive clade [group descendent from a single ancestor] than are chimpanzees or even bonobos." One telltale bit of evidence: Ardi lacks the fanglike canines that chimpanzees use as weapons.[5]

The common view of ourselves as warring predators looks at only a small fraction of our time on earth, after our numbers had grown large enough that we perceived our survival as dependent on territorial control. It ignores the vastly longer experience of our evolving species: that we developed who we are in close-knit communities, knowing our lives depended on one another. So deep did Charles Darwin perceive our connection to be that he surmised that primal people judged things good or bad "solely as they obviously affect the welfare of the tribe."[6] And Adam Smith—the very same philosopher who supposedly endorsed the "invisible hand" of selfishness—observed that love for ourselves is completely dependent on community.

> It is the great precept of nature to love ourselves only as we love our neighbor; or, what comes to the same thing, as our neighbor is capable of loving us.
>
> —Adam Smith,
> *The Theory of Moral Sentiments*, 1759

We're so socially constructed that isolation actually makes us sick. Being without friends, a Harvard University study of women found, is just as damaging to health as smoking.[7] "The subjective experience of social distance," says Steve Cole, author of a University of California, Los Angeles, study, "reaches down into some of our most basic internal processes—the activity of our genes." The study found that "the genes of chronically lonely people showed over-expression in immune system activation (such as inflammation) but under-expression in antiviral responses and antibody production," which help keep us well.[8] So, it is not surprising that solitary confinement is viewed, and experienced, as the harshest of punishments.

Flowing from this deeply social nature of ours, I'm convinced, are at least six human traits we can count on as we make this historic turn. As Adam Smith pointed out in 1759, "Man who can subsist only in society was fitted by nature for that situation for which he was made."[9]

## SIX HUMAN TRAITS WE CAN COUNT ON

But before I launch into my list, a clarification is in order. I sometimes hear "hardwired" used metaphorically to refer to our genetic endowment, but "soft-wired" seems more apt to me. Much of the expression of our genetic potential depends on triggers in our environment. This understanding, that context shapes the expression of an organism's potential, is central to an eco-mind.

So, here are six essential traits ready for us to tap as we create environments that draw them forth:

## One: Cooperation

It turns out that cooperating and co-creating explain our evolutionary success just as much as competition does.[10]

Human beings learned in our early tribal experience—how we lived for most of our time on earth—that we thrive within communities that work for everybody. Humans are unique among animals in our "pervasive sharing" of food, "especially among unrelated individuals," writes Michael Gurven, a leading authority on food transfers among hunter-gatherers living as our early forebears did.[11] The rule, except in times of extreme deprivation, is that when some eat, all eat. And the most productive hunters share the most.[12]

So, perhaps it should come as no surprise that neuroscientists using MRI scans discovered that when human beings cooperate, our brains' pleasure centers are stimulated as when we eat chocolate![13]

And what were the evolutionary pressures that turned us into cooperators?

In 2009, University of California, Berkeley, anthropologist Sarah Blaffer Hrdy challenged the common view that our cooperative nature was largely shaped by the need to bond in order to fight our neighbors. No, she says, confirming what Ardi seems to tell us. Over most of the 200,000 years we Homo sapiens have been around, there were simply too few of us to warrant fighting over territory. Instead, our capacities to cooperate as a group, to empathize, and to imagine what others are thinking—all of these traits—probably are responses to the selective pressures of evolving in a cooperatively breeding social group.

While other primates generally don't trust others to care for their infants, humans turn to aunties, grandmas, and friends to help care for their babies from birth. With these "helpers," children have the "luxury of growing up slowly, building stronger bodies, better immune systems, and in some cases bigger brains," Hrdy surmises. She reports that "shared suckling . . . occurs at least occasionally in 87 percent of typical foraging societies." And in one central African tribe, someone other than the lactating mother holds a baby during 60 percent of daylight hours.[14]

It is this capacity for cooperation, Hrdy stresses, that most distinguishes Homo sapiens from other species. So, might we thank our helpless, demanding babies not only for our cooperative nature but for all it's made possible—including civilization itself?

Interestingly, and contrary to modern assumptions, are additional findings aligned with Hrdy's thesis. "For over 90 percent of our existence as human beings, we lived, almost exclusively, in highly egalitarian societies," say University of Nottingham Professor Emeritus Richard Wilkinson and colleague Kate Pickett.

Such indications of our cooperative nature remind me of the musings of biologist Marshall Sahlins: "So far as I am aware," he said, "we [Westerners] are the only society on earth that thinks of itself as having risen from savagery, identified with a ruthless nature. Everyone else believes they are descended from gods."[15]

But there's more to grasp about what makes us unique that "cooperation" doesn't quite capture. It's our capacity for "shared intentionality," write Michael Tomasello and Malinda Carpenter of the Max Planck Institute for Evolutionary Anthropology in Germany. While chimps gesture in order to get others to do what they want, only humans form "a shared goal to which they are all committed, and know together that they are committed, and then form shared plans to reach the goal." Simply put, we know how to get things done together. Chimp involvement in group activity is more individualistic, write the two psychologists, but even one-year-old humans "form shared goals and plans."[16]

These distinctions may sound subtle, but they carry huge implications for grappling with today's challenges: Not just cooperating but co-creating is our nature.

## Two: Empathy

Cooperation is made possible by empathy, and it, too, seems to be a capacity deeply carved—or "soft-wired"—into us. Empathy, in turn, is made possible by our being able to see from another's perspective—a neat trick that even babies as young as seven months achieve.[17]

We also see a hint of early empathy in the finding that babies cry at the sound of other babies crying but rarely at a recording of their own cries.[18] And there's certainly no reason to think we humans are less empathetic than rhesus monkeys. In one experiment the monkeys refused food (in one case, going without food for twelve days) if eating meant they had to hurt another monkey with electric shock.[19] And notwithstanding the myth that crowds always panic in emergencies, research suggests that at such moments we're more likely to die from our compassion—as we use precious time to help others—than from a competitive stampede to save ourselves.[20]

In my own family, I'll never forget my son Anthony's story of his then two-year-old Josie asking, "Daddy, would you cry?" Anthony would pretend to sob, and his daughter would rush to comfort him. For me, it seems obvious that Josie wanted to experience the pleasure of comforting her dad back to smiles. Sophisticated experiments may prove our innate empathy and our pleasure in expressing it, but Josie cinches it for me. When Josie was three, I watched her offer her frozen lemonade to her crying cousin, who'd spilled his: "To help you feel better," she said. "It's all yours."

All this brings to mind a study reported in the journal *Science* in 2008: Two groups are given a chunk of money. One group is instructed to spend it on themselves, and the other, to spend it on gifts. Afterward, those who spent it on others reported feeling happier.[21]

> Once basics are covered, psychologists report that "the most important factor determining happiness is our relationships with other people. Might that also be true collectively? What does that imply about our species' estrangement from the rest of the biosphere?"
>
> —David Loy, *Money, Sex, War, Karma*, 2008

But what's most telling here is that the results surprised the subjects. Absorbing the view that we're nothing but selfish materialists, apparently we doubt the joy we experience in giving—even though studies show that on average over 80 percent of happiness comes from relationships, health, spiritual life, friends, and work fulfillment, while only 7 percent is about money.[22]

In *On Kindness*, Adam Phillips and Barbara Taylor remind us that we evolved to find pleasure in being kind. Not only is kindness pleasurable, they write, but it, "not sexuality, not violence, not money—has become our forbidden pleasure. . . . Once we allow it as a pleasure it makes us more porous, less insulated and separated from others."[23] A 1990s study of over 3,000 people confirms the authors' observation. It found fully 95 percent of volunteers reported that after helping others, they felt what the researchers called a "helper's high"—an improved sense of well-being, both emotionally and physically, plus enhanced energy and serenity.[24]

## Three: Fairness

In striving to create a social context that elicits our best, we can also count on a sense of fairness. It lives within most of us, for we learned a long time ago that injustice destroys community, the bonds of trust on which our individual survival depends. Adam Smith grasped this truth, noting that it is injustice that "will utterly destroy" society. So, more than two centuries ago, he wrote that we are "in some peculiar manner tied, bound and obliged to the observation of justice."[25]

And here, too, in our sharp sense of fairness, we're not alone in the animal world: Capuchin monkeys in a 2003 experiment flatly refused rewards when they saw that other monkeys were favored with tastier treats for the same effort. (The scientists, though, couldn't seem to bring themselves to call it a sense of fairness. To them, it's "inequity aversion.")[26] Even vampire bats and ravens detect overeating cheaters among those gathering food and punish them.[27]

What scientists see as a revenge instinct seems to be part of our fairness wiring. We know that if consequences for unfair action aren't imposed, the behavior might worsen and break down group cohesion. One study found that humans are a lot more likely to find the courage to impose sanctions on a transgressor if we know that others are observing us. Might this suggest that we don't necessarily find it fun to play cop, but if we're being judged, we want to do our part to protect the community?[28]

Revealing the effect of knowing others are watching, one study in a university setting found that even a mere wall photo of human eyes looking down on an "honor-system" coffee station meant users coughed up *three times* more for their drinks![29] In other words, we humans know what fairness requires of us, and being reminded that our social standing depends on being fair, we're more likely to act.

In fact, so committed are we to the principle of fairness that we will accept less for ourselves if that's what it takes to keep things fair. In a simple experiment, psychologists make subjects choose between the chance of getting nothing for themselves and the chance of getting less than a fair shake. It turns out that at least half of us will walk away with nothing before letting the other guy get away with treating us unfairly.[30]

Plus, fairness seems to make us feel good, even when at our own expense, *Nature* reported in 2010. In a simple experiment, pairs of young men were given $30 a piece, while at random one in each pair got a $50 bonus. For those receiving the bonus, the brain's reward center responded. No surprise. What was a surprise is that when these lucky men were then asked to imagine how they would feel if they got another bonus or if instead the bonus went to their partners, it was the second scenario, the one reducing inequality, that lit up the brain's pleasure center. "There could be an evolutionary preference for fairness," observed Elizabeth Tricomi, one of the psychologists involved, because it "helps us work together."[31]

These findings suggest that inequality in our societies, if perceived as unfair, abrades the human spirit and even harms the human body. In a rather stunning finding, British professor Richard Wilkinson and his team discovered that death rates among all fifty states do not correlate with rather large differences in average state-by-state income. Instead, they found a "strong relationship between death rates and income inequality" *within* each state. Inequality, Wilkinson finds, explains as much as a fifteen-year longevity gap between the poorest and richest groups. He goes to great pains to make clear that these differences are independent of, for example, better medical care for the better off. Rather, early death reflects the stress of "feeling looked down on . . . and being treated as second rate."[32]

## Four: Efficacy

We're also problem solvers. Think about it. Could our species have made it to 7 billion if we were essentially couch potatoes, shoppers, and whiners? I don't think so. We are doers.

"Man cannot tolerate absolute passivity," wrote social philosopher Erich Fromm in *The Heart of Man*. "He is driven to make his imprint on the world."[33] Our need to make a dent in the wider world is so great that Fromm argued in a much later work that we should toss out René Descartes's theorem "I think, therefore I am." Fromm sums up the human essence this way: "I am, because I effect."[34]

The trait seems to show up even in tiny babies. Three-month-olds respond with pleasure to a moving mobile, but, a study now shows, they "prefer to look at a mobile they can influence themselves." Plus, "they smile and coo at it more too," writes Professor Alison Gopnik in *The Philosophical Baby*. For her, the finding suggests that even the youngest among us enjoy making things happen and seeing the consequences.[35]

One aspect of being a doer, not a passive recipient, is the pleasure we experience in being helpful. And toddlers "have an almost reflexive desire to help, inform and share," reports psychologist Michael Tomasello, mentioned above, of the Max Planck Institute for Evolutionary Anthropology in Leipzig.[36]

Plus, little kids are smart about it. A study by Tomasello and colleague Felix Warneken measured toddlers' willingness to help an adult with tasks like stacking books, retrieving an out-of-reach object, or removing something from a box. Over and over, they helped the adult within seconds and without being asked. Apparently able to discern the need for aid, they didn't bother to help if objects had been deliberately discarded or books intentionally knocked over.

The investigators' conclusion?

That "even very young children have a natural tendency to help other persons solve their problems, even when the other is a stranger and they receive no benefit at all."[37] Tomasello concludes that "there is very little evidence . . . that children's altruism is created by parents or any other form of socialization."[38]

"Altruism" is probably a great descriptor, but I also see it through the lens of power—the deep need to master tasks and to be useful. Not long ago, walking through the supermarket behind my stepdaughter, I watched her two-year-old son, Ben, become quite annoyed when she didn't give him complete command of their shopping cart. Yes, he was helping out, but his deeper desire seemed to be to accomplish a difficult challenge.

In a widely known experiment carried out in the 1970s, Harvard psychologists Ellen Langer and Judith Rodin divided nursing home residents into two groups. In one, residents were allowed choices as to where to receive visitors and when to watch movies; they were also given houseplants to care for. Residents in the second group did not make these decisions and were told that the staff would care for the houseplants.

After a year and a half, the Harvard investigators returned and were surprised that fewer than half as many residents in the more engaged group had died compared to those in the more passive group. Langer attributes the stunning difference to the enhanced "mindfulness" of those making more choices.[39] I see the outcome differently. For me, the longer lives of those with responsibility for themselves and another living organism, the plant, affirm what Fromm so plainly states: We thrive when we know we have an effect, when we feel useful and that we count.

Interface, the eco-responsible business icon mentioned in Thought 1, reports that it gives employees at all levels of the company a framework for suggesting solutions to help move toward its environment-friendly goal.[40] Enabling its workforce to feel effective toward a shared higher purpose is likely one reason for the carpet company's success. And going beyond this "sense" of shared purpose to the fact of shared ownership is likely one reason cooperative enterprises are spreading globally.

*And power makes a difference.*

As our societies increasingly tell us that we are only responders to market forces and to distant, centralized corporate and government power, we feel less and less useful and effective. We are deprived of a sense of agency.

And when we *do* feel empowered, what can be the consequence? Let me tell you of an encounter I'll never forget that hints at an answer.

Almost a decade ago, my daughter, Anna, and I visited the homes of formerly landless workers in Brazil. They had gotten land of their own through the Landless Workers Movement (known by its Portuguese acronym MST), a nearly thirty-year struggle that has settled a third of a million families on almost 20 million acres of unused land. The MST is now arguably the largest social movement in this hemisphere. Landless workers, long seen as among the most powerless people in Brazil, have created thousands of new communities, farms, enterprises, and schools.

When the MST farmers explained to us that they were creating the first organic seed line in Brazil, we asked, "Are you moving to organic farming because as landless workers you experienced pesticide poisoning and now want to protect yourselves?"

I was surprised by the look on their faces. *You just don't get it!* they seemed to be telling us. "Our concern is to protect the consumer," they said.

Knowing all they'd been through, though, I heard even more behind their words: Why would we risk our lives (for over 1,000 MST members have been killed by landowners' hired gunmen and corrupt law enforcement) and work so hard to create this community, only to produce something (pesticide-laden food) that might hurt someone else?

Reflecting on this encounter, something clicked for me: As long as we feel like cogs in someone else's machine, we can tell ourselves we're not *really* responsible for the impact of what we do. But when people gain a sense of control over their lives, they're able to acknowledge the implications of their actions and feel good about taking responsibility. So, the obvious answer for our societies is—*no more cogs!* And in an ecological worldview of mutual influence, there are none. We realize we are co-creating our world moment to moment.

If Fromm is correct and our need for efficacy runs this deep, the challenge to our species is clear: Can we consciously reframe our need for agency in ways that align with the laws of life, with life itself? Can we shift from control as the primary expression of power and experience power as co-creating with others and with wider nature?

I'll return to this thought at the book's end.

## Five: Meaning

We human beings are creatures of meaning, seeking ways to give our days value beyond ensuring our own survival. This yearning for transcendent meaning runs as deep as any biological urge, certainly. (If you Google "God" and "sex," it's true that the later turns up more hits, but God is not far behind!)

Even the private act of voting isn't just about calculated self-interest, it dawned on me recently. It, too, is about meaning. Rationally, I can easily see that my single vote isn't about to decide anything. But entering the voting booth, I feel something more going on inside: a quiet sense of pride welling up because I know I'm playing my part in a larger human drama. So the effort we make to vote may be driven less by anything tangible we're voting for or against and more by the meaning we ascribe to common decision making itself—to the democratic ideal we're honoring by our act.

Appreciating that humans are motivated at least as much by the desire for meaning as by the expectation of material reward turns out to be critical in effectively encouraging eco-friendly choices. Consider this experiment in which subjects were asked to read about recycling: For one group of subjects, the task of recycling was framed as a way to save money; for the second group, the task was framed as benefiting the community. Those who understood the goal to be about benefiting others learned more deeply the content covered in the text, the researchers found, and were more likely to voluntarily visit the library and a recycling plant to learn more about recycling.[41]

So, in facing the environmental and poverty crises, we can count on our deeply human need to feel that our lives count for something big, and we can emphasize that one way human beings over eons have met this need is by striving to be good ancestors—enhancing our children's and their children's futures. Journalist Mark Hertsgaard ends his 2011 book *Hot: Living Through the Next Fifty Years on Earth* with a letter to his five-year-old daughter. His work suggests to me that we can each find meaning in thinking about what we will say one day when our children or grandchildren ask us, "Granddaddy, Grandma, what did you do to stop climate change?"

## Six: Imagination and creativity

Just as I thought the book was complete, it hit me: While *EcoMind* is all about the human capacity to rethink our world, imagination was missing on my short list of essential human capacities. Had I taken our imagination for granted precisely because it defines humans—it *is* us?

In building up to her discussion of human imagination in *The Philosophical Baby*, Alison Gopnik writes, "More than any other creature, human beings are able to change. . . . The key to human nature at every level from brains to minds to societies" is what neuroscientists call plasticity—"our ability to change in light of experience." And this quality, she underscores, depends on our extraordinary imaginations.

The great evolutionary advantage of human beings is our ability to escape the constraints of evolution, Gopnik reminds us. We can learn about our environment, we can imagine different environments, and we can turn those imagined environments into reality.

And it is the length of our protected childhoods, unique among species, that allows human imagination to flourish, Gopnik argues. Both "using tools and making plans . . . depend on anticipating future possibilities," and they play a "large role in the evolutionary success of *Homo sapiens*," she notes. And her experiments at the University of California, Berkeley, reveal these "abilities emerging even in babies who can't talk yet."

Gopnik invites her readers to consider the objects around her as she writes and to see that they are all, each and every one, products of human imagination. The point is so obvious that to make it stick, she gets dramatic: "I [referring to 'woman cognitive scientist'] am also a creation of human imagination, and so are you."[42]

Human beings' unique capacity for imagination ends this list because—coupled with our "plasticity"—it is what enables us to envision and make the changes we must in order to draw forth the *other* five essential qualities.

And it is this imaginative self that takes pleasure in the challenge.

## If we're so great . . .

If humans are all this—cooperative, empathetic, and sensitive to fairness— if we need to feel efficacious, seek meaning, and are naturally imaginative and creative, then why in the world do we mindlessly participate every day in a social ecology that generates so much destruction and misery for so many?

That's the big question.

Clearly with all the above going for us, human beings have what it takes to seize today's historic challenges. Yes, *but only if we get real.*

For me that means accepting the complexity of our nature and then, seeing with an eco-mind, fully appreciating the power of context— including conditions we ourselves create—to determine the qualities we express. To help me appreciate the power of environment to elicit expression of an organism's potential, I like to keep in mind the humble water flea. The size of a grain of rice, it's got about a third more genes than we humans do. While the water flea looks unassuming, in response to a threat in its environment, this tiny creature can actually grow whole appendages, its own "spear" and "helmet." From flea to human, organisms respond to stimuli.[43]

Now, back to what all this means to us.

Getting real starts with the admission that with the six magnificent traits above come some much less lovely aspects of being human. If we look at the grand sweep of history, it's impossible to deny that we—not a few but most of us—are also capable of unspeakable cruelty. In his book *The Lucifer Effect*, Stanford psychologist Philip Zimbardo reflects on his infamous prison experiment—a mock prison he set up in 1971 in Stanford's psych building. He divided young people who'd tested "normal" into prisoners and guards and dressed them for their parts. Zimbardo had designed the experiment to last for two weeks, but he had to call it quits in six days. The "guards" had begun tormenting their "prisoners" in ways gruesomely similar to what appeared in shocking 2004 photos of abuse by US soldiers at Iraq's Abu Ghraib prison.

Zimbardo's work, as well as earlier experiments by Yale University psychologist Stanley Milgram—not to mention what is evident in human horrors outside the lab, from the Inquisition to the Holocaust to slaughter in the Democratic Republic of the Congo today—reveal something about ourselves that is hard to swallow: that under certain conditions our pro-social qualities dissolve, and most humans will ruthlessly inflict pain on others. In his so-called obedience experiments, beginning in 1961, Milgram found that roughly two-thirds of subjects were willing to follow an authority's instructions even to the point of believing they were shocking a victim to death.[44]

Beyond this disturbing tendency to succumb to fear of authority or the desire to "fit in," even when that means going along with evil, another not-so-happy dimension of human nature seems to be a fairly universal tendency to experience pleasure when a rival falters. Apparently, empathy takes a holiday.

Germans call it "schadenfreude"—a satisfaction registered by the brain like the sensual pleasure of a good meal. English lacks a word for it, possibly because even acknowledging the feeling makes many people uncomfortable. Yet, psychologists have recently posited that its consequences for human society could be profound.

Because the thought of feeling glee at another's *undeserved* misfortune is so distasteful, we look for justification—we seek faults in the rival, and that faultfinding can spiral into group animosity. So, what might originate in a fairly innocent thrill at a rival's setback or a negative feeling about a rival's success, some social psychologists suggest, could swell to explain the deadly scapegoating of others among Hitler's disciples or Rwanda's genocide participants.[45] Yet another reason why social inequality—extreme and deepening, as in our society—is so dangerous: It means feelings of rivalry inevitably intensify.

Now, given our six deeply carved pro-social attributes, and accepting the ugly stuff too, *the* question for humanity seems relatively straightforward:

> *Which social rules and norms have proven to bring out the worst in humans, and which have shown to bring forth the best while protecting us from the worst?*

Here's my take. At least three conditions have been shown over our long history to elicit the worst in us, mainly because they deny so many the expression of our proven, richly positive capacities. These three conditions are: *extreme power inequalities*, showing up today in unprecedented economic disparities; *secrecy*, allowing us to evade accountability, as occurred when the financial industry, operating without transparency and public oversight, brought the global economy to its knees; and *scapegoating*, creating "the other" onto whom we can push blame, whether it's kids crying "he did it" on a playground or citizens at a town meeting shouting down a congressperson.

All three negatives seem to arise with ferocity in cultures premised on lack, where continuous rivalry is presumed. And, sadly, each has been on the rise in the US for at least three decades. Inequalities in income and wealth have reached or broken records, so today inequality here is much greater than in Ethiopia, to name just one country to which most Americans would not want to be unfavorably compared.[46]

Our soft-wired empathy and enjoyment in cooperation, our deep sensibility to fairness and need for meaning, efficacy, and creativity—all are stifled in societies where power is tightly held and opportunities shut off for so many. For me, it is not at all surprising, then, that scholars uncover a "strong relationship" between the extent of economic inequality and mental illness across countries.[47] This mismatch, between what we know brings out the best in us and the cultures we're creating, helps me understand why depression has become a global pandemic today: Worldwide, depression is the leading cause of disability, and the World Health Organization now ranks depression third as a contributor to the global disease burden.[48] More generally, the extent of poor health and social ills—from violence and crime to teen pregnancy and substance abuse—strongly line up with the extent of inequality across societies.[49]

So, when pundits in the US insist that our society's core problem is an intractable ideological divide—big-government believers versus small-government crusaders—or an unbreachable cultural divide—pro-life versus pro-conscience activists—let's not be confused. With an eco-mind we can stay focused on the deeper problem: the social ecology we're now creating that denies us the best in our species' own nature; a context in which we *will* point fingers—even point guns—at each other. In anger and frustration, we'll look for scapegoats and find them.

Knowing all this about ourselves, our challenge seems clear: How can we flip these three dangerous conditions by continually dispersing power, enhancing transparency, and assuming mutual accountability? In the process, we will be creating a culture of alignment with nature in which human needs are met in ways dissolving the presumption of lack. Throughout this book, I suggest that success requires rethinking, "reliving" democracy

WHAT WE NOW KNOW BRINGS OUT THE BEST IN US: IT'S MOVING . . .

from concentrated power to the dispersion of power;
from secrecy and anonymity to transparency;
from the blame game to mutual accountability.

itself. It starts with refusing to take our own positive proclivities for granted and instead actively nurturing them. All the great potential reviewed here needs our attention, and everyday living offers opportunities for its enhancement.

While most of us have experienced the helper's high and do have a Josie inside, we (and she, too) can also be cruel. Who hasn't cringed at the sight of kids taunting each other? Or felt repulsion when reading about teen suicides triggered in part by cyber- and text-message bullying?

But such cruelty is not inevitable.

All-important are the rules we humans make together. Vivian Paley, a seasoned teacher and MacArthur "genius," describes leading kinder-garteners in a discussion about the hurt created when one of them gets excluded from play. She asked the class to consider one simple rule—"You can't say you can't play"—which is also the title of her 1993 book. Several youngsters strongly resisted the rule, one even arguing that it would defy human nature to play with someone you don't like.

Gradually, though, the children came around, agreeing to live by the rule. She could almost hear a sigh of relief in the room, says Paley—perhaps arising from the children's freedom from the fear of being excluded them-selves. Years later, one of the students stopped Paley in a grocery store to tell her she's still committed to trying to live up to the "you-can't-say-you-can't-play" rule.[50]

This simple story is a powerful reminder that—notwithstanding hard evidence of our pro-social qualities—to really live up to the "best in us," we need context calling it forth.

In Norway, after three teenage victims of bullying killed themselves in 1983, schools galvanized to train everyone working with teens—teachers, janitors, and bus drivers—how to identify bullying and effec-tively intervene. Every week in their classrooms, children now participate in discussions about friendship and conflict. And parents are deeply in-volved. The payoff? Incidences of bullying dropped by half during the two years after these programs launched and remain low. Stealing and cheating also declined.

Those studying this success attribute it to the community's adoption of the values of dignity and respect. But my takeaway is that this "adoption" worked because of inclusive, deliberate, and specific actions—"shared intentionality"?—to create openness about the problem, along with new rules and expectations making *everyone* feel accountable.[51]

Our tendency to push blame onto others, intensifying in our culture, is not a fixture of the human experience. We have a choice.

Twenty years ago, farmers, loggers, and environmentalists—at logger-heads for decades in ongoing disputes over use of the land, how to respond to fires, and more—began to talk. Entrusted by the state to come up with a watershed plan, folks came together in the Applegate Partner-ship. Some got flack even for speaking to the "other" side. But members were determined to stop the blame game and created a button that members wore around town to symbolize their commitment. It was simply the word "they" with the universal diagonal line through it: No more "they." And with a lot of hard work, the diverse members of the Applegate Part-nership came up with a plan all could accept, protecting thousands of square miles of their beautiful state.[52]

Ending the cycle of blame also requires attention to nurturing empa-thy. Empathy is more than a "nice idea"; it's a "pragmatic skill." That's what University of Chicago professors Donald Scott and William R. Harper argue, as they teach empathy-based skills to their medical students.[53] Thus, we can deliberately develop these skills, striving to elevate their learning to a status as important as any tangible outcomes that families or schools or businesses generate.

The result can be powerful.

When students at the University of California, Santa Cruz, decided to launch a student-organized sustainability course, collaborating with the administration in order to green their campus, they realized their success would depend in large measure on how well they practiced what I call the "arts of democracy"—such people skills as active listening, mediation, ne-gotiation, and creative conflict. One source they turned to was Commu-nity at Work's *Facilitator's Guide to Participatory Decision-Making* and the training offered by this group. The students have succeeded brilliantly. Their course has spread to other University of California campuses and is touching the lives of thousands of students.[54]

Another art of democracy, perhaps the mother of them all—"civil courage"—I take up in the concluding chapter.

## SPEAKING TO OUR NATURE

So what might it mean to take to heart this understanding of our own nature as we work to make compelling and motivating the challenges before us?

In a recent TV interview with environmental luminaries, a fellow pan-elist declared that "people just aren't afraid enough. We have to increase the fear to get action on the environment." My heart sank.

We know too well that fear, particularly of our own death, typically brings out terrible things in human beings. Psychologists Tom Crompton and Tim Kasser report that most of us humans, when confronted with a survival threat, try to enhance our self-esteem through material gratifica-tion and by denigrating the "out-group," which in this case might well be environmentalists or even nonhuman nature itself.[55] We also try to avoid exposure to the bad news and to appear not to care.[56]

So, messages that trigger guilt and fear might enable the messengers to feel as if they're being tough realists.

*But are such messages actually tough?*

Not if they fail to challenge us to dig deep, to dig to the system roots of our crises. How often, for example, do I hear leaders bewail our lack of environmental progress and call for a string of desirable policies—from pricing carbon to upgrading public transportation—but then stop short, failing to name essential prerequisites, particularly political campaigns in which candidates are free from dependency on the very industries block-ing the policies we advocate.

*And are messages evoking fear and guilt realistic?*

Not if they end up backfiring because they don't incorporate what we now know about human nature—including how we typically react to threats.

Motivating each other using our eco-minds, we tell the truth but drop the scare tactics. We not only embrace but strategically use our deeply social nature. Sharing our knowledge of social and environmental threats, we emphasize stories about what *is* working with confidence that most of us are more likely to support an action if we know the ap-proach has already proven to make a difference. Respondents in one sur-vey were more apt to support climate-change laws, for example, if they were told a similar approach worked to confront acid rain. So, let's al-ways underscore the real-impact punch line in the stories of possibility that we share with others.[57]

Psychologists are also helpfully documenting that because our social nature encourages us to want to be like others, we're motivated by mes-sages that seem to offer us a way to meet that need.

In an experiment comparing the effectiveness of messages to encourage hotel guests to help save water and energy by reusing towels, psychologist Robert B. Cialdini found that imploring guests to the save the planet wasn't too effective. But this message was: "Join your fellow guests in helping to save the environment," followed by a note explaining that almost 75 percent of guests who were asked to reuse their towels did so.[58] Since it seems we're more likely to act when we believe others are, one easy takeaway for me is even more energetically to spread stories of everyday people in action.

Even more broadly, Cialdini warns that we make a big messaging mistake if we mainly scold those causing environmental harm—driving their SUVs, failing to recycle, leaving lights ablaze in empty rooms, and so on. The trouble is, what the public takes in is a message about "what others around us are doing," he says, and our "primitive" desire to be with the group responds.

His advice?

Heap attention on those people who *are* doing the behavior we seek. Cialdini tested his theories using a public service announcement in Arizona communities that combined these three messages: The majority of Arizonans approve of recycling, the majority of Arizonans do recycle, *and* they disapprove of the few who don't. These messages resulted in what Cialdini called an impact "unheard of" for public service announcements: a 25 percent increase in the tonnage of recycled material in communities exposed to the ads.[59]

To use our social nature strategically, we also can incorporate evidence that people tend to give more to a cause when their contribution is publicly visible—more evidence that our status with others matters enormously. So we can make more transparent all the ways people are contributing to social betterment.[60]

Finally, we can shed the assumption that in any simple sense humans don't like change. Knowing how much change must happen, and really fast, that thought is a killer. Sure, we do typically experience change in part as loss, but a striking feature of our species is our *attraction to the new*. Virtually from birth, humans are learners, testers, explorers. Even very young babies get bored with an "old" sound and, when hearing something new, "become attentive and start listening," says cognitive scientist Alison Gopnik.[61] So we can avoid any hint that environmental progress means return to a bygone era and celebrate humanity's fascination with the new, as do the many breakthroughs in this book.

I'm sure these findings only scratch the surface of how we can use our understanding of our social nature to become more effective communicators.

Most broadly, though, is this single key to effectiveness: We humans respond more powerfully to emotional than reason-based appeals, as psychologist Drew Westen underscores in *Political Brain*. Taking Westen to heart, we can deliberately speak to and seek to evoke—whether with the public, among friends, or in our own inner dialogue—a range of positive emotions about facing our global challenges. Why not these five?

> *Exhilaration* in feeling powerful as a contributor to something truly historic.
>
> *Dignity and self-respect*, for don't we all secretly want to be heroes, at least to ourselves?
>
> *Camaraderie* in knowing that we're walking shoulder to shoulder with others in common work.
>
> *Excitement* in novel experiences as we try out new ways of living.
>
> *Anger* at the needlessness of deepening suffering around us—a positive, too, if we have a framework for putting our anger to work.

So maybe it's time that we, whether of the Left or Right, stop cajoling others to be better people. The challenge is not *instilling* empathy or the need for a larger meaning in life or other pro-social qualities. They are in us. The real challenge is creating social rules and norms eliciting these qualities and actively nourishing them. It is making this century's planetary turnaround an epic struggle for life so vivid and compelling that it will satisfy these deep needs in billions of us.

As we strive to put forth an exhilarating ecology of hope, we can take heart in realizing that, like the formerly landless Brazilians, we are motivated to think through the consequences of our acts when we feel a sense of personal power. Thus, effective environmental messages must communicate a sense of possibility to be not just victims—or worse, mere perpetrators—of the crises but powerful contributors to solutions.

## THOUGHT LEAP 4

Sure, we can be selfish, fixated on material gain, and narrowly competitive. But here's the key to our future: We've also evolved deep capcities for cooperation, empathy, fairness, efficacy, meaning, and creativity. We can't change human nature, but that's OK. We *can* change the norms and rules of our societies to keep negative human potential in check and to elicit these powerful, positive qualities we most need now. Let's focus there with laser intensity.

But this new frame, which assumes we are able to choose new rules and norms to bring forth the best in us, runs smack into the next disempowering assumption about our nature.

thought trap 5:

# To Save Our Planet, We Have to Override Humanity's Natural Resistance to Rules

*Because humans—especially Americans—naturally hate rules*
*and love freedom, we have to find the best ways to coerce*
*people to do the right thing to save our planet.*

KNOWING HOW AMERICANS FEEL ABOUT FREEDOM, PRESIDENT George W. Bush found twenty-eight ways to get the word into his second inaugural address. The word "liberty" made it in fifteen times.[1] And while philosophers may debate its many nuances, there's no mistaking what freedom means to a lot of Americans: "Get out of my way!" So, it's no surprise that former Bush advisor Grover Norquist chose to name the allied conservative groups he has led the Leave Us Alone Coalition.

To love standing on one's own with unlimited choice—that's the definition of freedom Americans absorb. Public rules—laws and regulations—only limit and burden us. Not only do they violate our freedom, but for the most part rules don't work. Because people love freedom, they find ways around them. Rules must be kept to a bare minimum.

But wait.

Isn't it widely accepted that little kids need rules to feel safe and loved? Perhaps it's also true for the rest of us. In fact, it could be that unboundedness—having endless choice—is what makes us a little nuts. Rules and boundaries, spoken and unspoken, give our lives shape and structure.

Rules also offer meaning and a sense of purpose and connectedness to others—think of the Ten Commandments or the Bill of Rights or wedding vows. Writing this section on a Sunday morning, I flipped on the radio as a church service was ending: "You can take comfort in God's rules," said the minister, "because there are no exceptions."

Every sort of human activity—from marriage to mergers, from driving to dancing, from bed making to baseball—involves rules. Even little tikes are sensitive to playing by the rules—and the need to punish those who don't. "Toddlers," found a recent study, "selectively avoid helping those who cause or even intend [turns out they can tell!] to cause others harm."[2]

So, in many ways, perhaps all humans, even Americans, love rules. If true, the challenge of reversing our planetary downward spin is not to overcome the American character or human nature but to build on the natural human love of rules. In any case, whether we love 'em or hate 'em, rules are part of life: All systems, biological or social, operate by principles that determine what choices are possible for any participant in the system.

Since the nature of nature is rules, we can't do away with them, but what our species, uniquely, *can* do is make conscious choices about which rules shape our choices. We can make essential, urgent, social, and political changes so that we can create good rules, those that effectively serve life.

## Cues from nature

To go beyond a self-defeating antipathy to rules, we can take time to cele-brate that, unlike many human-made rules open to endless debate, nature offers us nonarbitrary, infallible guidelines. If we break the rules, there are heavy consequences, with no exceptions. Two obvious examples come to mind.

Humans turn up the concentration of atmospheric carbon by almost 40 percent and the consequence: a greenhouse effect. As a result, within this century, we face the spread of tropical disease into temperate zones, accel-erated decimation of species, flooding of coastal cities, and so much more.[3]

The rule in this case is simply that earth's atmosphere can absorb only so much carbon and still maintain the climate that has produced life as we've known it. So, taking our cues from nature, we can align hu-man rules with nature's: For example, tax what is ruining us—such as using too much carbon—which Finland began twenty years ago.[4] And we can couple that with lightening taxes on household income.

This latter rule change, dubbed "tax shifting," has had major ripples. Recall the Swedish town of Kristianstad from Thought 3, which freed itself from fossil fuels for virtually all its heating. Fueling this revolution was Sweden's carbon tax shift, subtracting on average $500 annually from each household's income tax and adding it mainly to vehicles and fuels.[5] In a related step, Denmark chose to tax fossil fuel energy in order to fund green energy, and today this tiny country boasts the world's tenth largest installed wind power.[6]

In the US, Boulder, Colorado, in 2007 put in place the first US tax on car-bon emissions from electricity. (Did you know that electricity generation in the US emits more $CO_2$ than all transportaton?[7]) And the city is using the revenue for its climate action plan.[8]

A second, obviously critical, rule of nature, ultimately determining whether we eat or not, involves the process of creating topsoil and the time needed. A widely used estimate is that one inch of soil is formed in one hundred years; yet, in the US, we've been eroding topsoil ten times faster.[9]

But humans are now learning how to re-align with nature's "soil rules." It starts for a lot of us with a big shift in perception: from soil as an inert medium for holding up plants to soil as a living organism with which we can co-create.

As explored in Thought 3, farmers in every corner of the world are relearning how to abet soil formation by returning organic matter to the soil and how best to mix crops to improve overall soil health. We're discovering how to vastly enhance fertility and reduce erosion by balancing soil microorganisms.[10] (Note that one gram of soil—well less than a teaspoon—can contain 10 *billion* microorganisms.) Farmers are reducing topsoil loss through "conservation tillage" on roughly 250 million acres worldwide.[11] From ancient farmers, we're also learning how to smolder organic matter, producing "biochar," as mentioned in Thought 3, that can enhance fertility by removing carbon from the atmosphere and can lock it away for centuries.

What comfort there is in this perspective on rules. We can take a deep breath, as we come to know that in aligning with nature's rules, we have something real to count on.

## RULE-MAKING POWER: WHO, HOW, WHY

A key to experiencing the relief that comes with rules that make sense and work is getting clear about what rules are now throwing us off. And that means letting go of the fantastical notion that the market, for example, operates automatically without rules: by "the magic of the market," as Ronald Reagan put it. The market embodies, and functions according to, vast numbers of rules that determine variables ranging from how wealth is distributed to how contracts are enforced.

So, the critical question for life is not whether there are rules (there are) but *who* makes them, the *process* of forming and changing them, and what *purpose* they serve. To step up to today's huge environmental and human challenges, we need to make sure that each of three elements of rule setting conform to what we know about conditions bringing forth the best in us. I've proposed that they include the dispersion of power, transparency, and mutual accountability.

We live in a society in which the corporation plays a powerful role in rule setting. Keep in mind that if we compare corporate sales to national gross domestic product, it turns out that just fifty-one corporations are bigger than half the world's one hundred largest economies.[12] So it seems smart to start with an understanding of the corporation: What is it, what is its role in shaping a wide range of rules determining our well-being, and what is its purpose?

OK. What *is* the corporation?

At our nation's founding there were only about six of them—public inventions created to serve a "public purpose" for a defined period, as, for example, building a road or bridge.[13] But the notion of a corporation as created by, and serving, society took a big dive in the late 1880s. It has not yet recovered.

In 1886 the Southern Pacific Railroad sued Santa Clara County over taxes it believed it did not owe. Ultimately the Supreme Court decided the case, and the published proceedings began, as was customary, with remarks prepared by a Court "reporter" (not to be confused with a journalist), in this case a former railroad president. He included comments that Chief Justice Morrison Waite had delivered before the court began hearing the case: "The court does not wish to hear argument on the question whether the provision in the Fourteenth Amendment to the Constitution, which forbids a State to deny to any person within its jurisdiction the equal protection of the laws, applies to these corporations. We are all of the opinion that it does."[14]

In a clarifying memo, Chief Justice Waite later wrote to the court reporter, "We avoided meeting the constitutional question in the decision," the question being whether the Fourteenth Amendment applies to corporations.[15] Nonetheless, the reporter's note prevailed. And the notion of "corporate personhood was established—without argument, without justification, without explanation, and without dissent," writes UCLA professor of law Adam Winkler.[16]

Strange, how history is made.

Winkler goes on to note that while corporations have been accorded some of the protections of real people, the record has been nuanced. In 2011, even a corporate-personhood-friendly Supreme Court unanimously rejected AT&T's claim that its personal right to privacy would be violated if records in a case involving the company's alleged overbilling were disclosed. "We trust that AT&T will not take it personally," wrote Justice John Roberts.[17]

Notwithstanding Roberts's lighthearted tease, the history of corporations accruing protections that were meant for real people is deadly serious. Since the late nineteenth century, the one body in our society most endowed with ultimate rule-setting and rule-dissolving power, the Supreme Court, has effectively put in place rules affording key rights, including "equal protection" and "freedom of speech," to nonhuman entities, rules that now arguably affect every aspect of human life and nature. If that claim sounds hyperbolic, consider that in early 2010 the conflation of a person's and a corporation's rights of free speech (plus defining spending as a form of speech) culminated in a momentous 5–4 Supreme Court decision: It overturned one hundred years of precedent and unleashed limitless corporate and union spending in political campaigns.[18]

Four out of five Americans disagreed with the Court's decision.[19] They could no doubt see that it further favors large corporations' power over citizens' power within our political system.

My hunch is that most Americans would agree with Chief Justice John Marshall, who in 1819 wrote that the corporation is an "artificial being."[20] Common sense tells most of us that corporations cannot share the same rights with us because they are incapable of bearing the public responsibilities of real persons. They cannot vote or sit on a jury. Corporations are not able to pledge allegiance to the US because their allegiance is instead to shareholders, who could live anywhere with interests opposed to the national interest. Other pertinent differences with real people, with huge consequences, are these: The human beings behind corporations are in many ways legally shielded from responsibility for their actions (thus the term "limited liability"), and since their resources can become vastly greater than those of mere mortals, they can wield those resources to bend public choices to serve their private interests.

These and other differences between corporations and people have enabled them to wield enormous influence in political rule setting. The consequences of the growth in power of corporations, increasingly protected as "persons," combined with the notion that markets work on their own, have been staggering.

## Market mythology, market meltdown

In the 1990s the financial industry grew so fast that, by the 2000s, its share of corporate profits had jumped to 40 percent, double that in the mid-1980s.[21] With public oversight considerably reduced, the industry took on risk, leading to global financial meltdown. Yet the financial industry's power remained so great that it could rebound to record profits in only a few years, even while the wider economy had regained only one in nine jobs lost.[22]

This massive increase in human suffering flows from the wild notion that we humans function best with minimal rules—an idea promoted religiously in schools of business and Washington think tanks—and the faith that, on their own, says Eugene Fama of the University of Chicago's business school, markets are rational, and prices are the "best estimates of intrinsic values." Nobel Prize–winning economist Paul Krugman notes that in recent decades these views have gained virtual cult standing among many economists.[23]

In fact, it was a weakening of the rules, via the financial services "modernization" act of 1999 signed by President Bill Clinton, that in part made possible Wall Street's stampede to the ethical bottom. It removed barriers to company consolidation and conflict-of-interest regulations.

Knowing this history, blaming "Wall Street greed" for the financial crisis misses the real lesson: that it is the *rules and norms we as a society put in place* that determine how most of us behave.

Mike Francis, an executive at Morgan Stanley when the mortgage meltdown began, described how mortgage standards sank and sank—and ultimately hit bottom with what brokers dubbed "NINAs," the no-income, no-asset loan. In comments broadcast on National Public Radio, Francis called it "a liar's loan," acknowledging that "we are telling you [the mortgage seeker] to lie to us. . . . We're setting you up to lie. Something about that . . . felt wrong way back when, and I wish we had never done it, [but] we did it because everyone else was doing it."[24]

When the context—the rules and norms we create—allows it, most of us are apt to go along with the crowd against our better judgment, just as Francis did. That's the downside of our social nature. And when that context includes secrecy, the outcomes can be really bad.

At the height of the shenanigans that brought on the collapse, observed University of Texas government professor James Galbraith, some Wall Street players lived by an insider "code": IBG-YBG, an acronym meaning "I'll be gone. You'll be gone."[25] In other words, since we're operating in secrecy, we're invisible. So when it all implodes, we can just slip away unseen. And most did.

Unfortunately, such a lack of transparency is almost certain to arise where power is tightly held.

Even former Federal Reserve chair Alan Greenspan, once an acolyte of antigovernment author Ayn Rand, has come to see the risk of too-tightly-held power. He argues that for a market system to work, rules must prevent power from becoming too concentrated—to keep companies from becoming "too big to fail." If they are that big, "they're too big," Greenspan said in the fall of 2009. One answer is enforcing the antimonopoly rule on the books for a century. "In 1911 we broke up Standard Oil—so what happened?" Greenspan asked rhetorically. "The individual parts became more valuable than the whole. Maybe that's what we need to do."

Soon, though, Greenspan backed away, but we don't have to.[26]

Absorbing the Great Recession's excruciating lesson in what happens when we believe that rules are expendable, can we step up to keep the market open and fair? The answer will depend on whether we elect to office those who understand that rules are not anathema to a vibrant market; they are essential to it. And for that shift to become increasingly possible, we can change another rule: We can remove the power of private wealth in campaigns for public office.

## ONLY CHANGE IS CONSTANT

But, I hear a reader whispering, how can we even begin to imagine re-embedding such powerful entities as corporations into an ecology of democracy—shaped by a transparent, accountable set of rules beholden to our well-being?

It helps me to keep in mind that throughout most of human existence, economic life—making and trading to meet our needs—has not been a separate realm of life at all, and certainly not one above all others, but embedded in family, culture, and ritual.[27] Today's corporate dominance is an aberration. With an eco-mind, we see the corporation being shaped moment to moment by our ideas, laws, and court rulings, as well as by our actions as purchasers, workers, and citizens. They, too, are participants in an ecosystem.

To see the reality of continuous change inherent in an ecological worldview, we might start by rejecting any notion that what I call the "one-rule economy"—that is, business driven only by highest, short-term gain to top executives and shareholders—has always been the American way. It has not. A team of distinguished economists writing in the 1950s observed that US executives believed they had "four broad responsibilities: to consumers, to employees, to stockholders, and to the general public."[28] Even as recently as 1981, the Business Roundtable, representing CEOs in many of the biggest US companies, stated that a corporation "must" consider its impact beyond shareholders to "the society at large."[29]

"The corporation is an evolving entity, and the end of its evolution is by no means in sight," said President Harry Truman's advisor Edward S. Mason in 1968.[30]

## POSSIBILITY

No law requires a corporation's myopic focus on maximizing immediate shareholder return, or at least claiming to, even when executives are the big winners. Corporations are chartered by state statutes, and no state instructs directors only to consider *short-term* shareholder interests. That pressure comes in part from the threat of hostile takeovers if stock prices dip and from executives' fear of losing their jobs or not collecting their juicy bonuses.

In fact, during the last two decades, roughly 30,000 companies have joined some sort of business organization setting standards for more responsible behavior—now a $40 billion market.[31] Even at the very center of corporate business training—Harvard Business School—Professor Michael Porter, chair of the school's program for new CEOs of very large corporations, now calls for a new form of capitalism. He's christened it "shared-interest" capitalism, in which corporations operate on the premise that their success depends on society's advancement. "The purpose of the corporation must be redefined as creating shared value," he and coauthor corporate-responsibility specialist Mark R. Kramer write, "not just profit per se."[32]

Alone, voluntary initiatives can't get us where we need to go, but they are changing expectations and can become part of a transition returning the corporation to its original role—that of an agent of public progress, subject to rules set in the interests of thriving societies.

And it's possible. State statutes governing corporate charters, it turns out, aren't set in stone.

Responding to heightened threat of hostile takeovers, starting in the 1980s, a number of states amended their corporate statutes to permit corporate boards to consider the impact of their decisions on broader constituencies, such as workers and the community, when making merger-related decisions.[33] More broadly, in Pennsylvania, for example, directors are explicitly not required to put the interests of one group— even shareholders—first.[34]

Skeptics may laugh—since none of this *requires* corporate boards to behave differently—but what these changes prove is that a corporation's marching orders, the statutes defining it, aren't diamond-hard.

And that's where Jay Coen Gilbert comes in.

Three years ago, the lack of universal corporate standards—with teeth—taking into account firms' total impact on people and the planet became too much for Jay. So in 2006, he and colleagues launched B Corporation (*B* for beneficial), based near Philadelphia, to create a certifying process through which over four hundred companies, including King Arthur, the country's second-largest flour maker, have reworked their corporate charters to "bake their values into the DNA of their company," says B Corporation.

The charter of a B corporation *requires* it to make "a material positive impact" on employees, the community, and the environment and to make publicly known its "social and environmental performance" as measured by third-party standards.[35] By 2011, four states—Maryland and Vermont, New Jersey and Virginia—had passed legislation enabling a firm to incorporate as a B corporation. Similar laws are making their way through several other state legislatures.[36] Philadelphia views B corporations so positively it even offers them special tax breaks.[37]

Take heart, also, that most US companies are *not* giant multinational corporations traded on the stock market. They are privately held and family owned, which gives them a great deal of flexibility. They produce half of private, nonfarm gross domestic product.[38] This big group can't rationalize an immediate-profits-only mind-set by claiming that the rule of the market or a charter obliges companies to look out only for shareholders.

And get this. It turns out that social responsibility of the sort required by the legal structure of the B corporation leads not only to community and environmental benefits but to better financial returns as well: *On average socially responsible firms perform better financially than their less responsible competitors.*[39] Even during the financial meltdown, companies prioritizing sustainability did better than their counterparts.[40]

## Good-bye to the fiction
### that corporations are persons

Someday, with the hindsight of history, we may look back and see that right now, out of sight of most of us, a most critical step is advancing the goal of embedding rules of economic life in a relational, ecological worldview. For example, citizens are starting to say, *No, the notion of corporate personhood is not working.* More than that, they are helping make new rules, democratizing the rule-setting process.

By 2009, over a hundred US municipalities, mostly in Pennsylvania, were creating new rules to challenge corporate rights. They have passed ordinances asserting community rights to self-governance and banning corporations from harmful operations, such as factory farming and spreading sewage sludge.[41]

In 2009, citizens of three Maine towns—Shapleigh, Newfield, and Wells—resisted the mining of their groundwater by Nestlé (for its Poland Spring bottled water) by declaring that within their boundaries, corporations are not persons. Shapleigh resident and business owner Shelly Gobeille, leader of Protect Our Water and Wildlife Resources, told Nestlé Waters North America that she was "outraged" that Nestlé "came into my community without permission or even the knowledge of residents."[42]

Gobeille went door to door to let her neighbors know. And before it was all settled, citizens helped to pass ordinances that also included the right of ecosystems to flourish and the right of citizens to self-governance—including the right to place groundwater in a public trust.[43]

The Nestlé Corporation backed down.

And lest anyone assume such a radical step might be possible only in small communities, look at what Pittsburgh did.

In 2010, Pittsburgh's city council approved an ordinance elevating "the rights of people, the community, and nature over corporate 'rights'" in its successful strategy to prevent natural gas drilling—shown to contaminate groundwater, among other dangers—within city limits. Councilman Doug Shields defined the battle as deeper than drilling itself: "It's *about our authority as a community to decide*, not corporations deciding for us."

In their strong stand, Pittsburgh leaders have stood up to the gas industry in ways Congress has not: For a quarter century, pressure from the industry has thwarted Washington lawmakers' efforts to regulate hazardous waste, reports a 2011 *New York Times* investigation. In 2004, the EPA studied the type of drilling proposed in Pittsburgh, ignored warnings in an early draft, and asserted that drilling "poses little or no threat." Says EPA whistle-blower Weston Wilson, most of the study's review panel were current or former oil and gas employees.[44]

The Pittsburgh ordinance includes a local "bill of rights" that covers the right to water and to local self-government. Moreover, with this ordinance, Pittsburgh became the first US city to recognize the legal rights of nature.[45]

## WILDFIRES OF COMMON SENSE

Letting go of any assumptions about human beings' knee-jerk antipathy to rules, we realize that new social rules, aligned both with nature's nonarbitrary laws and with our own nature, can take shape and spread quickly if they make sense to people—if new rules and norms ring true to us and if we feel engaged in shaping them.

Herb Walters, founder of RSVP/Listening Project based in North Carolina, explained to me recently that many conservative Christians have had negative perceptions of environmentalism because it's been linked with liberal ideology, secularism, and "environmental extremists." They have felt none of this included them.

But a Listening Project in two rural North Carolina counties found that resistance often morphed into action when hour-long interviews with trained listeners offered church leaders a chance to reflect on stewardship of the earth as God's creation and, from there, to reclaim their biblically based sense of responsibility for creation care. A number then formed a faith-based organization that's effectively taken on protecting the watershed, helping connect low-income homeowners with a community agency offering weatherization and launching countywide sustainable economic development.

> Activists often see themselves as having the answers. Listening Projects change the rules by treating each person interviewed—even our opponents—as the expert. When we listen to people, they may find common ground or offer solutions. Some then take the next step, working for positive change.
>
> —Herb Walters, founder, RSVP/Listening Project

The Listening Project approach, triggering profound changes on a range of concerns nationally and internationally, is both simple and profound: When we're listened to, we start to make connections. New rules that honor those connections start to make sense.

And once a change of rules makes sense to people, it can take off with amazing speed.

Consider what I call the "Reward Renewables Law" in Germany (officially, the Feed-In Tariff), described in Thought 3. Taking effect about a decade ago, it rewards households producing renewable energy. In just a few years, the simple tool has already spread to over eighty-two countries, states, provinces, and municipalities.[46] Lightning speed.

A recent adopter is Gainesville, Florida, where households are charged about thirteen cents a kilowatt-hour for electricity. Now, however, their publicly owned utility will pay *them* thirty-two cents a kilowatt-hour over twenty years for any solar-powered electricity they generate. (Those who come on board after two years will get somewhat less but will still benefit.) And all the while, they'll also enjoy knowing they are not contributing to toasting our planet.

Because it made so much sense, the initiative moved ahead quickly, with a unanimous vote by the Gainesville City Commission.[47]

Other cases of speedy takeoff in response to new rules?

Partly in response to the state's rebate incentives, the number of California's solar-powered rooftop installations grew from 500 to 50,000—that's *one hundredfold*—in the last decade alone. Ten states doubled their rooftop solar capacity in just one year, 2008.[48]

Another sign of people's willingness—even eagerness?—for structures guiding and amplifying their action is the response to a 2005 campaign by former Seattle mayor Greg Nickels. Frustrated by the Kyoto Protocol–resistant Bush administration, he began reaching out to other mayors to entice them to pledge to greenhouse gas emissions reductions faithful to the protocol. Nickels hoped that, maybe, he could bring 141 cities on board—his magic number because it is how many countries had signed on to Kyoto.

But six years out, *over 1,000* mayors from the fifty states, the District of Columbia, and Puerto Rico have signed on to the Mayor's Climate Protection Agreement.[49]

After Supervisor Ross Mirkarimi in San Francisco sponsored a 2008 bill banning plastic bags from supermarkets and drug stores, he got calls from city officials all over the country wanting to follow suit. Mirkarimi exclaimed, "I think we ignited a wildfire of common sense!"[50]

It happens.

## Rules that work

In our society, it's frequently assumed that in economic life government rules burden businesses, which then pass on their costs to us. So rules may be necessary, but they are expensive.

What this view skirts, first, is how expensive is the *lack* of rules. Recall the cost of mining and using coal in the United States: one-third of $1 trillion in environmental and social costs every year. Then imagine what government rules moving us away from coal could save us.

Some rules offer immediate savings.

New fuel-efficiency standards that went into effect in 2010 will save us a bundle. Saving starts immediately, and assuming gas at $4 a gallon, in ten years consumers will be saving $58 billion annually, nearly $200 for every US man, woman, and child.[51]

Or think of what Filipinos are saving because their country put in place a no-incandescents rule. In 2008, with help from the Asian Development Bank, the Philippine government gave out 13 million compact florescent lightbulbs for free and then phased in, over two years, a ban on incandescents. The move is cutting household lighting costs by as much as 80 percent, while reducing greenhouse gas emissions.[52]

To further free ourselves from the dangerous assumption that rules hinder and restrict progress within the drip, drip, drip of antigovernment rhetoric over thirty years—recently becoming a Tea Party torrent—it helps simply to remind ourselves of the wide benefits we experience today because of public rules put in place over this same period.

From California to Massachusetts, here are just two vivid examples touching me personally.

In 1982, a pretty simple rule changed the way the Public Utility Commission of California rewarded power companies: They were allowed to keep as extra profit part of whatever it saved its customers. Soon the companies began investing hundreds of millions in efficiency. "It's cheaper than building power plants," said Arthur Rosenfeld, former member of the California Energy Commission.[53]

People's energy bills fell significantly, and utility shareholders enjoyed greater dividends.[54] Now, a handful of states, including Vermont, Idaho, New York, and Hawaii, have followed suit.[55] The approach is likely one reason California's per person use of electricity has hardly changed in a quarter century and is about 60 percent of the national average, while its greenhouse gas emissions are roughly half the national average.[56]

Long before this shift, when I was a kid in the 1950s visiting my Aunt Vivienne in Los Angeles, my eyes would burn from the smog. But in the early 1970s, the Clean Air Act and state measures kicked in, and by 1996 Los Angeles had cut by 94 percent the days each year that air exceeded the "Stage 1" smog alert level compared to 1975.[57]

Now, living in Boston and driving along the Charles River, I am continually reminded of the powerful role that public rules, along with determined citizens, can have in rescuing nature's gifts. By the mid-1960s, the picturesque river was in sad shape. Sewage from outdated waste-water treatment plants flowed into it, along with industrial toxins that turned its water pink and orange. As the river became known for fish kills, dumped cars, and bad odors, no one dared swim in the Charles or eat its fish.

But a citizens' watershed protection group jumped in to save the river, using the 1972 Clean Water Act to give it clout. This led to new waste-water plants and strict limits on industrial discharge. State rules and public works helped, too, including big sewer improvements that greatly cut raw sewage. In 1995, the Environmental Protection Agency prioritized making the Charles safe. Today, the river is swimmable much of the year.[58]

These are very specific examples of human-made rules effectively confronting flagrant disregard of nature's rules. Unfortunately, however, given the extent of private influence in Washington, what I call our "privately held government" is often blocked from preventing violations of even the most critical of such laws: The *New York Times* reported in mid-2009 that polluters have committed more than half a million violations of the Clean Water Act since 2004.[59]

But even when enforced, such case-by-case efforts that limit harm can only take us so far.

## Commons care

Although dramatically improving our lives, the Clean Air and Clean Water acts are nonetheless reactive. To protect what we love, we can dig deeper. We can dig to the underlying ground rules that shape our behavior and economic life more broadly. As we strive for a truly publicly held government, accountable to us, we can work to avoid harm, not just limit it.

One approach is the framework I think of as "care for the commons"—those precious, indivisible goods such as air, water, soil, forests, oceans, and diverse species that we inherit, share, and yearn to pass on unharmed or enriched to our children.

When as a young woman I first began thinking about the "big questions," the whole idea of the "commons" had just been slammed. A framing metaphor capturing a lot of imaginations at the time was the "tragedy of the commons," the title of a 1968 *Science Magazine* article by Dr. Garrett Hardin. What stuck was the idea that, because each of us is motivated to pursue our immediate self-interest, anything held "in common" gets trashed—even though the effect is to mess the whole thing up for everybody.

The idea grabbed people, I think, because it fit into the bigger frame that "humans hate rules," including rules that could protect the commons from precisely this tragedy. But Hardin's metaphor, however memorable, is mistaken. Sure, instances abound of exactly the behavior he spoke to—like today's massive ocean overfishing. But the classic "commons"—grazing animals in seventeenth-century England, for example—generally worked well, as do many modern examples. In Törbel, Switzerland, communally owned grazing meadows, forests, irrigation systems, and roads all function well and have for at least five centuries.[60]

Then, history in our thinking about the commons was made again: More than forty years after Hardin's powerful metaphor, Dr. Elinor Ostrom received a Nobel Prize in economics for showing the conditions under which commons *do* work: when participants make and enforce fair rules for their use. Certainly, we know that a commons fails—as global climate chaos proves—when power imbalances are so extreme that it's not possible to hold each other accountable. Part of the problem is that when a small minority is in control, the rest of us can feel "off the hook" and fail at the basic human task of protecting what we love. (A great place to keep learning about the commons is at OntheCommons.org.)

> We come to value what we share as much as what we own, what keeps us alive as much as what we exchange.

The point of commons care is to prevent harm before it occurs. And an eco-mind can help, shifting our way of seeing from the dominant way of thinking about solutions—doing less harm to something outside of us—to that of alignment within our ecological home. We come to see natural treasures no longer as merely divisible property but as gifts protected by boundaries we create and honor, knowing that all life depends on their integrity.

In this shift, we come to value what we share as much as what we own, what keeps us alive as much as what we exchange.

## Saving the sea

Internationally, an example of a potentially useful framework of rules for commons care is the UN Convention on the Law of the Sea, covering everything from navigational rights to conservation and management of marine resources, with disputes settled through binding arbitration by the International Court of Justice.[61] Since 1994, 161 nations have ratified it, but not the US. We're the only holdout among industrial countries. Joining us are Iran, North Korea, Cambodia, Libya, and a handful of other countries.[62]

But as ocean champion Dr. Sylvia Earle reminds us, even with the treaty, almost two-thirds of the ocean is beyond the claims of various countries and remains a "great blue free-for-all."[63] Clearly, the treaty's effectiveness isn't impressive.

So far.

But more and more of us understand that typing up rules on pieces of paper can't get us there; we need the active support and "buy-in" of people everywhere. That buy-in may become easier as more of us see the cost of *in*action mounting, with massive assaults on ocean health having wiped out 90 percent of large predator fish and damaged vital coral as well as phytoplankton. The latter aren't exactly a household term—yet— but these microscopic plants form the basis of the entire oceanic food chain, perform half the earth's photosynthesis, and therefore provide the earth half of its oxygen.[64]

We can urge Congress not to allow a handful of opponents in Congress to block ratification of the convention. After all, the second Bush administration called ratification "urgent."[65]

Humanity has brought about 12 percent of the world's land area under common protection but less than about 1 percent of our oceans.[66] This might change once the fact sinks in for more of us that phytoplankton, like forests, are essential carbon absorbers in our fight against climate change.[67] A sign? In 2010, Canada made history in creating the world's first conservation reserve that reaches from mountaintop to ocean floor— over 2,200 square miles that include part of coastal British Columbia so rich in life it's dubbed the "Galapagos of the North."[68]

## Forest guardians experience immediate gain

Back on land, new forest commons are emerging too.

In Ecuador, for decades indigenous people have fought to keep oil companies from despoiling rain forests there. And in 2010 they finally won. The UN Environment Program and the government of Ecuador signed a first-in-the-world agreement to keep oil reserves untouched forever beneath the Yasuni National Park—a rain forest that's almost the size of Connecticut. In exchange, the world's nations—because we'll *all* benefit—are establishing a $3.6 billion trust, to be forfeited if the oil is ever drilled. Ecuador gets to use the interest earned, about 7 percent, or $252 million each year, for conservation and green energy, as well as for health and educational programs benefiting the indigenous people.[69]

So, if you're one who doubts human capacity to think long term, as any commons requires, let this sink in: In a poll before the Yasuni deal was settled, three-fourths of Ecuadorans opposed oil drilling, effectively relinquishing a near-term financial gain of $7.2 billion in revenue.[70] Recall, also, Costa Rica's similar choice not to exploit its oil.

Preserving *all* the world's rain forests would cost about 0.1 percent of the value lost in the 2008 stock market collapse, an environmental advisor to Prince Charles has calculated.[71]

Seems doable to me.

Rules for commons care involve a lot more than big-time treaties and reserves.

In India, for example, it's what neighbors take on together. Since the 1990s, villagers have been cooperating to ensure their forests' regrowth. Today about 10 million rural households take part in roughly 100,000 "forest-management groups." Each creates and enforces rules to prevent overuse of nearby woodlands. Motivation is high, especially for women, because firewood still provides three-fourths of the energy they use for cooking. So, groups with a larger proportion of women—a quarter or more—"have done particularly well in improving forest condition," reports economist and environmentalist Dr. Bina Agarwal.

The forest-management groups, working in collaboration with the Indian government, cover a fifth of India's forests, and they're likely a reason that India is one of the few countries in the world to enjoy an increase in forest cover since 2005.[73]

On the other side of the globe, Costa Rica fosters commons care with another strategy: rules rewarding landowners who maintain forests, which both absorb carbon and protect the watershed for the benefit of fisherpeople and farmers. In 1997, the country created another rule, a carbon tax of 3.5 percent on fossil fuels, and began using the income to pay these forest protectors—now numbering about 7,000. Another rule, a water tax on big users, like hydroelectric dams, farms, and suppliers of drinking water, goes to pay villagers for keeping rivers clean.

Besides being "a major source of income for poor people," the approach "has also enabled Costa Rica to actually reverse deforestation," says Carlos M. Rodríguez, the country's former environment minister.[74] In the mid-twentieth century, Costa Rica's rate of deforestation was among the worst in Latin America. But just since the 1990s, its forests have spread at an amazing speed, from only covering a fifth to now covering half the country.

## The rights of nature inseparable from human well-being

Digging even deeper, consider another Latin American country's bold step in 2008 to redefine its relationship to the natural world. Ecuador became the first nation to affirm in its constitution that "la naturaleza también tiene derechos"—that nature has rights too. Indigenous Ecuadorians helped to craft the new constitution, which allows citizens to bring legal cases on behalf of the natural world.[75]

Then, in 2009 Bolivian president Evo Morales called on the UN General Assembly to develop a Universal Declaration of the Rights of Mother Earth. And in 2010, almost 70 percent of Kenyan voters approved a new constitution that recognizes the right of every person "to a clean and healthy environment, which includes the right . . . to have the environment protected for the benefit of present and future generations."[76]

Here at home, but unbeknownst to most Americans, more than twenty US communities have conferred legal rights on nature.[77]

## "Future Justice" incorporates the well-being of life still to come

As part of this global movement for commons care, five years ago, Hungarian environmental lawyer Sándor Fülöp, fifty, became the country's—and the world's—first parliamentary commissioner for future generations. He served as environmental ombudsman for the environment, the legal representative of generations to come. In this role, his office weighed in on big legislative matters ranging from dangers posed to residents by a proposed military radar station to a threat to historic vineyards from an oversized biomass plant. Fülöp was happily surprised by the support his new office got from other branches of government. One observer credited his success to his ability to engage with all the stakeholders involved.

In New Zealand, the parliamentary commissioner for the environment—empowered to "investigate and report on any matter where the environment may be, or has been, adversely affected"—is known as the "Guardian of the Long View."[78]

Simultaneously, the Hamburg-based World Future Council, of which I'm a member, has put forth the concept of "Future Justice"—adding to the world's jurisprudence a legal obligation to those who follow us.[79] The council is creating tools for identifying policies most damaging for future generations and also loopholes in current laws that can be closed to protect them. It will also shine the light on exemplary rules and policies—in part via a Future Policy Award—that safeguard life for those to come.

Might these historic shifts portend humans' diminishing sense of separateness from nature? It's true that many US environmental laws, like the Clean Air Act, consider nature as mere property, in effect legalizing environmental harm by establishing permissible levels of pollution and destruction. But, in giving nature legal standing, we move from trying to effect less damage toward a different relationship with nature altogether—a relationship of mutuality, not dominance.

A big leap, yes—but, again, recall that our own Declaration of Independence grounded our right to liberation from British rule in "the Laws of Nature." *So, if nature can confer rights on us, might we come to see the logic of acknowledging nature's rights too?*

## REAL FREEDOM

As long as we think of freedom as the absence of rules, we can't know freedom, for we position ourselves outside of nature—including our very own. The question for humanity is not about more or fewer rules. It is whether we can together create rules that serve life because they align with all we know about what makes us thrive. In part, that means envisioning economic life as just one dimension of an "ecology of democracy" in which rules keep the market open and fair, and the corporation returns to its original function in the service of community well-being.

We can then experience real freedom: not freedom from rules but freedom with power—freedom to participate in creating rules that promote life.

### THOUGHT LEAP 5

The nature of nature is rules, and humans love rules that enhance our sense of belonging and give our lives structure, shape, and meaning. They work and take off fast when people feel engaged in shaping them, and they make sense to us. Knowing all this, we can go beyond rules that limit harm and establish ground rules that avoid harm to begin with.

All that I've argued here depends on our species being able to experience itself as connected with, and part of, the natural world. But some people understandably fear it's too late to reawaken those feelings.

thought trap 6:

# HUMANS HAVE LOST THE CONNECTION TO NATURE

*Now thoroughly urbanized and technology-addicted,*
*we've become so disconnected from nature that it's pretty hopeless*
*to think most people could ever become real environmentalists.*

I T CAN CERTAINLY FEEL THIS WAY. AFTER ALL, HUMANITY DID TURN A BIG corner in 2008. As of that year, the United Nations says, more than half of us now live in cities.[1] In the UK, the typical eight-year-old is better at recognizing Pokémon characters than common wildlife.[2] And I heard recently of a teacher whose grade-school students believed buffalo wings come from, yes, buffalos with wings!

Now only 6 percent of American children nine to thirteen years old play outside unsupervised with other children as they have from time immemorial. "I like to play indoors better, because that's where all the electrical outlets are," a fifth grader told Richard Louv, author of *Last Child in the Woods*.[3] And I myself realized I had to change the cricket ringtone on my iPhone, lest my Manhattan-bred granddaughter, who at three may not yet have heard a real cricket, someday encounter the sound and begin scouring the grass for someone's lost phone.

Given these huge changes, maybe it's too much to expect people to feel themselves truly part of nature, sharing one fate, and therefore naturally eager to reverse the human assault on nature.

But wait. *What if we just can't help it?*

Whether we consciously feel connected or not, human beings have been soft-wired over eons of evolution to love and respond to nature. There's even a fancy term for it—*biophilia*—first used by Erich Fromm but expanded by famed Harvard biologist E. O. Wilson to describe the connections that human beings subconsciously seek with the rest of life.[4] A few urbanized decades and technology immersion can't knock it out of us.

Because of our innate connection, almost any exposure to nature, studies show, seems to do great things for us—benefiting our health in many ways, including helping us to recover from stress and to concentrate and think more clearly. And what's dawning on more and more urbanites, as well as planners and developers, is that city living doesn't have to break this natural connection.

Most obvious is the enticement of the garden. Roughly a third of American households—41 million—gardened in 2009, up about 14 percent in just a year.[5]

There are now roughly 18,000 community gardens in the US and Canada, including those on otherwise vacant lots and land in public housing projects.[6] And all of these little plots add up. By one calculation, between 15 and 20 percent of food produced in the world is grown by 800 million urban and peri-urban farmers and gardeners in cities from Havana to Kampala.[7]

Sometimes the connection with nature comes as a big surprise to people. "You mean, all this time I have been hungry and have sometimes had to go without food, and now I find out food grows in the ground?" asked a resident in the garden at Interfaith House in Chicago.[8]

It's also worth noting that the United Nations estimate with which I began—that now more than half of us are urbanites—may overstate our distance from the natural world. Typically counted as "urban" is any population of 2,000 or more; yet, when I lived in a Vermont town of 5,000, I was surrounded by hundreds of forested acres, and my life felt deliciously rural. So, most of humanity still has wide access to nature.

## SOFT-WIRED CONNECTION: THE PROOF IS IN

Now, let these findings assuage remaining fears that humans have lost an innate connection to the earth:

Just a view of a park or greenery from home has a positive effect on the cognitive functioning of children.[9]

The same findings hold in the workplace. "Over long periods people working in rooms with windows have fewer illnesses, feel less frustrated and more patient, and express greater enthusiasm for work."[10]

Before and after showing subjects three hundred photos of urban and rural scenes, University of Essex researchers put them on a treadmill. In subjects who had viewed pleasant rural scenes, they discovered, blood pressure fell significantly, while it climbed in those who'd seen the unpleasant urban photos.[11]

And when we're ill, nature helps us heal: Two groups of patients had the same surgery in a 1984 study at a suburban Philadelphia hospital but experienced very different outcomes. The difference depended on whether their rooms looked out on a brown brick wall or a stand of deciduous trees. Following their surgeries, patients who had the tree view experienced shorter stays (nearly one full day on average) and needed less pain medication. Their moods were better, and they suffered fewer minor complications.[12]

Benefits to patients occur even if nature is reduced to a recording or painting. In a Johns Hopkins Hospital study, one group spent the hours before surgery listening to recordings of birdsong and a babbling brook. They could also look at a large landscape picture. The control group had no picture or sounds. The "nature" group reported significantly better pain management than the control group.[13]

In the Netherlands, 1,000 farms are revealing the healing power of growing food. On these "care farms," farmers are paid to incorporate into the workforce people suffering from mental illness and addictions, as well as young people who've been dismissed from regular schools. The results are impressive—as significantly fewer of these folks later need social-services help. Farmers benefit too, reporting that they love sharing their passion for farming, feel less isolated, and enjoy the "meaning brought to their lives through delivering care."[14]

Living near a park or other green space boosts health and encourages health-related behavior so much that UK scientists calculated that the health gap between rich and poor might be cut by half through exposure to green spaces.[15]

In general, "evidence suggests that children and adults benefit so much from contact with nature that land conservation can now be viewed as a public health strategy," argue environmental health advocates Howard Frumkin and Richard Louv.[16] We might "prevent or treat illness," Frumkin suggests, "by prescribing gardening or pet ownership or vacations in beautiful places."[17]

Now that's health-care reform!

And doctors are catching on. In Santa Fe, New Mexico, doctors are combating runaway diabetes by distributing trail guides to patients in a program called "Prescription Trails."[18]

## HUMANS MEET NATURE, EVEN IN CITY CENTERS

We may not be able to turn the city into a Garden of Eden—which was, it's useful to note, organic, with a wild diversity of plant life—but urban dwellers are discovering ingenious ways to meet their inborn need to connect with nature.

Few would have guessed Chicago would take the lead. Since 2001, its city hall boasts a garden roof with 150 plant varieties covering a full city block. The garden's insulating effect saves the city $5,000 a year on utility bills.[19] Green-roofed city buildings are sprouting from Seattle to Austin, and this movement is global. Atop São Paulo's city hall is a lush garden. In New York City, as in cities across the globe, rooftop gardens grow not only decorative plants but also food, offering fresh produce for individuals, restaurants, and even schools.[20]

And for many, it's food, glorious food, that awakens a whole new awareness of our connection with earth.

It sure worked for me.

Today, more and more eaters are joining this revolution of the senses, becoming producers themselves. Not only are urban dwellers increasingly growing at least some of their own food—adding to the 800 million of us worldwide—but novel partnerships are forming.

At the Milk and Honey Café in Philadelphia, owner Annie Baum-Stein took the worldwide decline of bee colonies to heart and in the spring of 2010 asked her neighbors and customers if they'd like to be part of her summertime "host a hive" program. Thirty-five people responded, and Annie began with seven new homes for bees—three rooftops (including the store's) and four backyards. All the hives were successful—but they brought some surprises, too: The taste of honey from the store's roof and honey from just four blocks away turned out to be delightfully different.

And in one host family, young kids created a PowerPoint about their hive to share at school. They "sent me pictures of the presentation," Annie said, "that brought me to tears of pride." Annie's now planning to continue "host a hive."[21]

Connecting eaters and growers on a grander scale is community-supported agriculture (CSA), taking off in less than three decades. From a single CSA farm in Massachusetts in 1986, now over 4,000 dot the country.[22] The model is simple: Farmers sell "shares," collected before the growing season from future eaters of their crops. They then don't have to take out interest-laden loans, and eaters are assured great food during harvest season.

Madison, Wisconsin, may be the CSA capital of the country, with nearly forty-six such farms nearby. A favorite memory is visiting Vermont Valley Community Farm CSA just outside Madison with my daughter, Anna. We arrived on a Sunday afternoon to discover a pesto-making party under way. In a huge barn, member families were busily blending freshly picked basil and garlic with olive oil. The memory of the mouth-watering aroma of their creations is still with me.

For me, part of the brilliance of the CSA model is that it embodies "mutual accountability" so critical to the broader positive social and

environmental transformation under way. It is one of the three conditions I posit throughout this book as essential to progress: We stop
simply blaming, we drop victimhood, and we step up together as co-
problem solvers. In a CSA, eaters share the inherent risk in farming with
the producer.

In my family's CSA, Waltham Fields Community Farm near Boston, we
love the pick-your-own moments when we get to search for the very best
strawberries or tomatoes. Picnicking Saturdays at tables near the fields,
we delight in watching the little kids, including four-year-old Ben (remember the shopping-cart commander in Thought 4?) wander the rows.
The children seem to connect with the earth immediately. Recently, my
son, Anthony, told me about his family's visit to Red Hook Farm in Brook-
lyn, where three-year-old Josie went right for the earthworms and was
enthralled. And Ben? He announced that for Halloween he is going to be
a farmer!

Farmers' markets are taking off across the Western world. In the UK
their earnings more than doubled in two years. Here, too, the number is
burgeoning, from under 2,000 in the early 1990s to over 6,000 in 2010.[23]
People seem to love not just the high-quality produce but also the convivi-
ality the markets bring out: Shoppers engage in ten times more chatting at
farmers' markets than they do in supermarkets, sociologists tell us.[24]

Local Harvest, a national good-food group, offers a nifty website to
make it easy to find a CSA, a farmers' market, or a community garden
nearby.[25]

Whole towns are getting into the food act.

Two hundred miles northwest of London is Todmorten, which a few
years ago tagged itself "The Incredible Edible Todmorten," and appropriately so, for not only does food now grow around its pubs, but raspberries
and apple trees adorn the local health center grounds, and the train platform is garnished with mint and rosemary. Even the town's Victorian
cemetery grows onions and strawberries. Posted around town are maps
showing twenty-six public gardens where anyone can harvest ingredients
to take home for dinner.

This burst of gardening passion in England is a nationwide affair, with
a 700 percent increase over twelve years in the waiting list for space at

community gardens. So a "land-share" website, now with over 40,000 users, serves as a nationwide matchmaker, linking the "land-but-no-time" Britons with those eager to get their hands in the dirt. And London's mayor pledged two years ago to create 2,012 new vegetable patches across the city by 2012.[26]

Some fear that this food revolution can only recruit from the better off, since too many low-income neighborhoods have become "food deserts"— deserted by all but convenience stores. Yet, striking energy and creativity is coming from folks hurting the most.

Take Oakland, California, where the poverty rate is 50 percent above that of the nation as a whole. A budding Food Policy Council—coordinator of innovations to make good food available to all—used GPS and aerial photos to help locate about 1,200 acres of urban open space, including over 750 publicly owned tax parcels. Most are near public transportation and schools. Organizers figure that the plots potentially could provide as much as 10 percent of Oakland's vegetable needs. Plus, urban agriculture, they note, offers jobs, green space, and educational opportunities as it contributes to public health and safety.[27]

In Indianapolis, the city Office of Sustainability—does your town have one yet?—has come up with a unique idea for reaping these benefits *and* helping solve another problem: property tax revenue lost when our economy tanked and residents had to abandon homes so dilapidated they had to be demolished. In 2010 the city began offering eighty-five plots to citizens for free, if takers commit to farm them for five years. The idea is that entrepreneurs—seven got started in the first summer—will produce food for their families or sell to restaurants and at farmers' markets.[28]

And Detroit, seen as down for the count by so many, has vast potential to experience a food-led turnaround. Recently, University of Michigan scholars documented that, by growing food on half the land that's now vacant and publicly owned, Detroit could produce 40 percent of the fresh, nontropical fruit, as well as two-thirds of the fresh vegetables, now consumed. And that's assuming low productivity. With high productivity, this land could meet the same share of the city's considerably higher recommended consumption of fruits and vegetables.[29]

Any city could try it.

## RETRIEVING LIVES THROUGH THE LAND

Growing Home, a social enterprise in Englewood on Chicago's struggling South Side, offers homeless, low-income people and ex-offenders (about 90 percent have been incarcerated) work in one of its three organic farms to gain skills and reenter the workforce.

The farms also produce healthy, organic food for the neighborhood and for top restaurants and farmers' markets as well.

Growing Home's executive director, Harry Rhodes, explains why the approach works: "Getting your hands dirty, planting, watching things grow and harvesting crops is very self-affirming."[30] Two-thirds of the folks who go through the seven-month program, covering farming basics to marketing skills to personal finances, get full-time jobs or pursue further job training. Nearly 90 percent are able to rent an apartment of their own or find other stable housing.[31]

Growing Home is showing how urban farming can change not only individuals but communities. Its Wood Street Urban Farm led to the founding of the Englewood Farmers Market, connected to an active neighborhood committee that's increasing awareness of fresh food and healthy eating. The next big step happened in early 2010: a Greater Englewood Urban Agriculture Task Force was born and has met monthly ever since. Over eighty stakeholders take part, from aldermen and city officials to urban-farm activists and small-business people, along with academics and potential funders.

"Our goal task is to turn Englewood from a 'food desert' to a 'food destination,'" Harry told me. That means thinking big for Englewood: expanding from a handful to a large number of urban farms; providing training to prepare farmers to run their own farm businesses; and helping birth new farm and food-related businesses in Englewood.

On the northwest side of Milwaukee, not far from the city's largest public housing project, is the heart of an even bigger food enterprise started in 1993 by six-foot, seven-inch former basketball player Will Allen. Called Growing Power, each year it produces healthy food—from duck to collards, from herbs to perch—that benefits thousands of area residents through schools, restaurants, co-ops, a farm store, and farmers' markets, as well as through low-cost CSA food baskets delivered throughout the neighborhood.

From its original two-acre site, Growing Power now works ten additional urban farms employing forty people with help from over 2,000 volunteers a year, including a lot of young people. Combined, these green oases grew enough food in 2009 to feed over 10,000 people.[32]

Growing Power has a really big goal: to "grow food, to grow minds, and to grow community." On any given day there, you can learn the ins and outs of composting, participate in a "From the Ground Up" workshop to learn how to develop and run food initiatives in your own community, or, if you're a young person, take part in trainings that range from agricultural basics to leadership and entrepreneurial skills.

On its initial two-acre site in Milwaukee, Growing Power packs in fourteen greenhouses, where herbs and vegetables grow almost up to the ceiling. Out back are pens full of chickens, goats, ducks, turkeys, and beehives.

Plus, an ingenious symbiotic system annually raises 10,000 tilapia and perch. Water is pumped upward for sprouts growing above a long tank. The sprouts' dirt and gravel filter the water, which trickles down, pure and aerated, to the fish.[33]

With big help from 250,000 worms, Growing Power each year also turns 12 million pounds of spoiled food, which otherwise would have rotted in city landfills, into rich compost and produces more than 10 million pounds of new soil. It's sold for income and also nourishes the soil on Growing Power's nearby forty-acre farm and in gardens throughout the city.[34]

"I am a farmer first, and I love to grow food for people," Allen says in a *New York Times* profile.[35] "But it's also about growing power." It all adds up to more than a movement, he says: It's a "good food revolution" that is truly multicultural. People of color and immigrants, some who used to look at gardening as slaves' work, he notes, are reconnecting with their natural environment to grow the power of their communities.[36]

## "Teach them to long for the endless immensity"

"If you want to build a ship, don't drum up people together to collect wood," French writer Antoine de Saint-Exupéry advised, "and don't assign them tasks and work, but rather teach them to long for the endless im-mensity of the sea."[37]

Many people are taking to heart Saint-Exupéry's insight, asking how to do more than "teach children about the environment" and, rather, instill in them a longing for nature's "endless immensity."

School gardens, in which kids plant, tend, harvest, and in some cases prepare the food, are catching on. California's goal is one in every school; already, a garden is part of roughly a third of the state's 9,000 plus schools.[38] My daughter, Anna, and I visited one in Berkeley where hay bales were interwoven with the rows of tomatoes, squash, and kale to make an outdoor classroom. We laughed in amazement when the teacher defended the choice of kale as one of the starter plants for kids to plant, cook, and serve. We assumed kale would be their last choice, but as part of a bruschetta dish the children were making, they loved it.

California may be the leader—certainly it's got a climate advantage compared to many states—but others across the country are making gardening part of schooling.

Richard Louv's 2005 bestseller *Last Child in the Woods* awakened many people to what he calls our "nature-deficit disorder." Since then he's launched a back-to-nature movement drawing on numerous studies showing that play in a natural setting increases children's motivation and problem-solving and cognitive abilities (generating a 27 percent increase in mastery of science concepts alone).[39] Children demonstrate more self-control and experience less stress, along with less attention-deficit disorder and obesity.[40]

The movement, sometimes called "No Child Left Inside," has caught fire in more than fifty regional and statewide campaigns. In 2008, the US House of Representatives passed the No Child Left Inside Act. If pushed forward and approved by the Senate, It would get environmental education back in the classroom—and by extension, kids back into the natural world.

Pressing forward with such initiatives, we can take heart from the story of our nation's great conservationist Teddy Roosevelt, a privileged city kid who ended up saving over 234 million acres of America's wilderness. It began when Roosevelt, a nearsighted child of twelve, was fitted with glasses and immediately fell in love with birds.[41]

And Teddy didn't even have the benefit of programs such as the "forest schools" spreading across Europe.[42] As part of a Forest Education Initiative in the UK, children spend entire days in the forest, helping to set the rules and exploring the forest together. In one exercise, five-year-olds are asked to collect items "that don't belong"—that is, which are human-made—along with their favorite natural items. They construct mobiles with them and then talk together about their preferences. Most of the children find nature's offerings the most appealing (though one lad preferred a discarded wrapper because he said it reminded him of his favorite "crisps"!).

Young people in the US and elsewhere are also taking the lead in "greening" their college campuses. As part of the Real Food Challenge, students ask their schools to pledge to ensure that 20 percent of campus food will be "real"—sustainable, fairly produced, and local—by 2020. The founders had hoped 200 schools would get on board in the first year, but they hit 329 instead.

As we think about and participate in these varied, life-changing initiatives emerging across our country and planet, let's be sure not to frame them as "going back to nature." "Going backward" is not what most humans find appealing, and the frame is false in any case, since we are never separate from nature.

Finally, to reinforce our confidence that a yearning for reconnection with nature is within most of us, let me share a story.

Recently, on a gorgeous day in the central Massachusetts town of Sturbridge, population 9,000, I learned about the Great Tomato Plant Giveaway and Gardeners' Fair. Residents gathered on the town common last spring, where twenty booths introduced them to the joy of growing, and children planted their own scarlet runner beans.

But here's what amazed me: Four hundred participants stood patiently in line, some for at least an hour, to get their free tomato seedlings—including heirloom varieties Purple Prudence and Brandywine—along with simple growing instructions. Organizer Ellen Duzak told me that someone asked how many years it would take to grow the first tomato crop. Yes, we can chuckle at how little most of us know about growing food, but we can also marvel at the strength of the desire tapped through this simple community event.

Thus, with mounting evidence that our bodies are designed to respond to the natural world and with ingenious strategies popping up, let's assume that humans can recall our natural bond with nature. And with that confidence, let's quicken the pace, as climate disruption insistently reminds us of the consequences of our forgetting.

## THOUGHT LEAP 6

We can count on one thing for sure: Humans evolved to love nature. So city dwellers, too, can and are rediscovering that natural connection, finding that it enhances our health, learning, and fun. As we reconnect with nature in all these ways and more, it's likely that support for public action to heal the natural world will grow too.

# thought trap 7:

# IT'S TOO LATE! ]

*It's too late! Human beings have so far overshot what
nature can handle that we're beyond the point of no return.
Democracy has failed—it's taking way too long to face the crisis.
And because big corporations hold so much power, real democracy,
answering to us and able to take decisive action, is a pipe dream.*

"TRYING TO SAVE THE PLANET IS JUST A LOT OF NONSENSE . . .
because we can't do it," says ninety-two-year-old Professor
James Lovelock, known for his "Gaia hypothesis" that the biosphere is a
single living organism.[1] Ross Gelbspan, acclaimed author of *Boiling
Point*, tells us that "we have failed to meet nature's deadline" and warns,
"The environmental establishment continues to peddle the notion that
we can solve the climate problem. We can't." Even if humanity tomorrow
were to reject oil and coal, "it would still be too late to avert major climate
disruptions," he argues. Sea level is rising twice as fast as predicted, and
drought is already killing crops. Famine, disease, displacement—all are
certain, he says.[2]

Much more often I hear, simply, "Yep, game's over. We're f—ked."

To those declaring our species' near-future demise, I find myself wanting to shout, *Wait a minute!*

Half the world is getting by right now on a daily sum equal to the price of a single American latte—or less. About 1 billion of us lack the food and water we need.[3] In the Global North, millions are struggling and stressed as well. Even before the Great Recession, it was estimated that almost 60 percent of Americans will live in poverty for at least a year during their adult lives.[4] In short, catastrophe is *already* the daily experience of huge numbers.

So here's my question: Too late for *what?*

I agree with Gelbspan that it *is* too late to prevent massive change in the climate we humans have taken for granted for thousands of years. Erratic, extreme, and destructive weather is already with us. It *is* too late to prevent suffering. Terrible suffering is already with us.

But it is not too late for life.

Life loves life. Life strives to "create the conditions conducive to life," says "biomimicry" champion Janine Benyus.[5] That's just what we do. In other words, the very essence of life, including the version we call human—creatures who yearn to make their mark in the world—isn't changed by climate chaos. It is not too late to be ourselves. In fact, for our species, with its passion for shared action toward common ends, maybe now is *the* time to be alive.

When facing staggering setbacks—from the Black Death of the fourteenth century to world wars killing nearly 100 million people in the twentieth—most human beings don't end up ruing life. What makes us miserable isn't a big challenge. It's feeling futile, alone, confused, discounted—in a word, powerless. By contrast, those confronting daunting obstacles, but joined with others in common purpose, have to me often seemed to be the most alive.

So it's not too late to take in today's explosion of learning about our own nature and about the nature of our earth and then to team up for a momentous passage to a world more aligned with our learning and with our deepest needs.

Neither is it too late to learn how to grieve our losses—to grieve, for example, the daily demise of entire species, most of which we've pushed out before we've even met them. Grieving is critical to staying alive.

So, we can stop worrying about whether it's too late—what a waste—and get on with the only thing we really should wonder about, even obsess over:

*What's keeping more of us from jumping into action? What's preventing us from being ourselves?*

Especially now that, with an eco-mind able to see our connectedness, we know that our actions *do* matter—virtually every single one of them—and they will determine not just the extent of the loss that our own and other species will experience but our chances of evolving ways of living together vastly more satisfying than those we know today.

So what *is* stopping us?

I've argued throughout that we've allowed a set of false but potent ideas to tie our hands, to make us believe that we are powerless. I fear that just beneath the "it's-too-late" lament—now rising in volume—lies a paralyzing despair fueled both by ideas and, in this case, a lack of them. True, I've challenged the premise of lack at every turn, but there is one lack we *do* have to worry about: Clinging to a failing idea of democracy, we lack a vision of a practical and powerful process bringing us together as the problem solvers we yearn to be.

A huge blind spot.

So, the "it's-too-late" thought trap is, for me, less about a trap and more about a deficit—a deficit of imagination. Whether and how we erase it will determine, more than anything else, the shape of life to come.

## THE MOST WORRISOME DEFICIT

We in the US, along with many others worldwide, seem to be suffering from what linguists call "hypocognition," the absence of a key concept a society needs to thrive. In our case, it's a missing concept of democracy effectively aligned with our real nature—the good, the bad, and the ugly; a democracy able to bring out the best in us and to protect us from the worst in us.

Collectively, we in the US have grown up with a dangerous notion of democracy as some*thing*, a fixed structure we've inherited—something done *to* us or *for* us. This understanding of democracy is so mal-aligned that it brings out the worst in us, while blocking expression of the best.

Given the shortcomings of what we've been calling democracy, some people I've long admired for their deep commitment to addressing our global crises seem ready to give up on it. They may not share the Far Right or Libertarian perspective that virtually all government is bad. But they see the US government, ensnared in a lobbyist's cage, as incorrigible. A hopeless cause. Some, like, acclaimed psychologist and author Daniel Goleman in *Ecological Intelligence*, seem to discount the role of government altogether in greening our economy, suggesting that with more knowledge of the impact of our shopping choices, our green purchases almost single-handedly could mold a sustainable economy. Such a narrow focus on our potential power as shoppers feels rooted in resignation to our meager power as citizens.[6]

Not long ago, when I was discussing the global crises with a small group of allies, one confessed that for him, China, totalitarian though it is, has an advantage when it comes to facing climate change. It can mandate solutions from the top.

I cringed.

My friend failed to note that China ranks way down the list in the Environmental Performance Index, at 121 compared to the US at 61, and China is still investing heavily in dirty energy, including one new coal plant a week.[7] But author and columnist Thomas Friedman laments that China seems to be moving faster toward a green economy than the United States.[8] Already, the world's leader in wind energy, China's plans to double its capacity in only five years.[9]

And my response to China's positive steps?

It's not to condone totalitarianism, which is profoundly mal-aligned with deep human needs. It is first to acknowledge that even in totalitarian China, where workers' wages as a share of GDP are falling fast, pressure from everyday people can matter. The World Resources Institute's Deborah Seligsohn notes that on the environment, decisions in China are not "purely top-down. . . . Everyone who comes here feels the bottom-up ferment." She adds that "the environmental ministry has actually used public opinion very, very effectively to support its role." The minister realizes "the public really cares about the environment."[10]

My second response is horror at the obstacle our thin *version* of democracy—what I call "privately held government"—poses to our progress here, driven as it is by private, not public, interest. And this is why I agree with Al Gore that "in order to solve the environmental crisis we have to solve the democracy crisis."[11] Even more emphatically for me, in so many ways the environmental crisis *is* the crisis of democracy.

## WHY DEMOCRACY NOW

What does it take, then, first to rethink democracy as a key to solutions and then to believe we're capable of it?

When we get *really* frustrated with our struggle for genuine democracy and are impressed, like Friedman, with some steps top-down China is now taking, it helps to remind ourselves soberly of what we know for sure.

Over and over, change makers who believed they alone held the answers have consolidated power into their own hands, only to end up perpetrating humanity's worst missteps. "Top-down" might *look* attractive till you think of where "I'm the decider" George W. Bush took us—to what could easily have turned out to be a multi-trillion war in Iraq with thousands of American deaths and, according to the journal *Lancet*, hundreds of thousands of "excess Iraqi deaths as a consequence of the war."[12]

So, we can drop the dangerous idea that centralized power—derived from the mechanical command-and-control view of life—works. The early-2011 mass resistances to its failures in the Middle East are only the most recent reminders. Contrary to lessons drummed into us, concentrated power is often not resilient, efficient, or smart.

> It's fair to say that nature abhors monopoly.
> —Jane Jacobs,
> *The Nature of Economies*,
> 2001

And it's not "natural." Think about it: What really messes with a healthy ecosystem is its takeover by a single invasive species—think kudzu spreading in the American South—whose dominance undermines the give-and-take of a complex and dynamic balance among diverse species.

Aside from the horrors unleashed when power imbalances are extreme, top-down approaches are doomed to let us down for other reasons.

One is the very nature of today's problems. They are complex, interrelated, and pervasive. They involve all of us. So answers require changes in all of us. But humans don't typically innovate by fiat. We change when we sense that something big and important depends on us and when we feel "ownership," knowing our voices and actions count. Top-down strategies fail to tap citizens' ingenuity, creativity, energy, experience, and diverse perspectives—everything we need right now more than ever.

Think of the Will Allens of the world, making urban landscapes edible and beautiful. Or of Jay Coen Gilbert coming up with a solution to the corporation's narrow logic? Or of those gutsy Niger farmers greening millions of acres, once a shift of perception empowered them to embrace and improve traditional approaches.

There's much more, though, to "why democracy now"—notwithstanding evidence from a range of countries revealing deep disillusionment with its failing forms.[13] To deepen our passion, we can let go of the idea that because democracy is hard for us, it must be unnatural. From a range of disciplines, evidence I noted earlier is mounting that human beings have just the essential capacities needed for real democracy.

And even beyond our species: While animal-behavior experts used to think that it was the dominant leader who made decisions for the whole herd, they're discovering that it doesn't always work that way. For instance, red deer, native to Britain, move only when 60 percent of the adults stand up. Whooper swans of northern Europe "vote" by moving their heads, and African buffalo do so by the direction of the females' gaze. Scientists also suggest that this sort of animal "democracy" may carry with it a real survival advantage.[14]

Looking at how humans perform in groups, a 2010 study reported in *Science* noted "converging evidence of a general collective intelligence factor." In almost seven hundred subjects working in groups of three to five, researchers discovered that better performance on a range of tasks is not "strongly correlated" with the average or maximum intelligence of participants. What matters, it turns out, are three factors: how many females are in the group, the "average social sensitivity" of participants, and the degree of "equality" in taking turns during conversation.[15]

Here's more evidence that as we genuinely listen to one another— what democracy is all about—we come up with better answers.

Also motivating for me is the realization that democracy is essential to keep wealth and its power widely dispersed, and economic inequality itself serves as an obstacle to our aligning with nature. For one, in unequal societies people feel less secure, and feelings of insecurity tend to drive us to focus more on the material—on purchases that can bolster self-esteem.

Plus, the more unequal societies are, with the US one of the world's most extreme, the lower they typically rank in measures of social and physical health and the greater their rates of violence and incarceration.[16] The recent global financial crisis was itself "driven by income inequality," argue two economists with the International Monetary Fund.[17] All these problems divert attention and resources away from the positive avenues to environmental wholeness celebrated throughout this book.

Thus, the cruel fact that two-thirds of US income gains between 2002 and 2007 went to the top 1 percent is, among other harms, a big *environmental* problem.[18] So "environmental groups [are] realizing that they've got to get out of their green silos," declares Naomi Klein, author of *The Shock Doctrine*, "and start engaging with economics."[19] Precisely. We environmentalists can show how hard economic facts impact ecological well-being and fortify our passion for democracy by pointing to real-life evidence linking democracy and ecological thriving.

## WHAT WORKS . . . WHAT DOESN'T

Costa Rica pops up as a global green leader throughout this book. Yet its income inequality—while less than in most of Latin America—is high. So what lessons does Costa Rica hold?

You may have heard that Costa Rica is one of the few countries without a military. So, the answer must be that with no military expenses, more resources are free to serve environmental goals. But this answer only raises another question:

Why is Costa Rica different?

Over the last one hundred years or so, Costa Ricans have come to understand the citizen not as "an ideal aspiration" but as "a key actor in practical politics," writes Professor Chalene Helmuth in *Culture and Customs of Costa Rica*.[20]

After its civil war ended in 1948, Costa Rica not only abolished the army but also passed universal suffrage and created the nonpartisan Supreme Electoral Tribunal to uphold the integrity of voter registration and election results. To overturn the tribunal's advice requires a two-thirds vote by the legislative assembly.[21] (Imagine if the US had been able to turn to such a tribunal in November 2000 to resolve our disputed presidential election.) The tribunal "has all but eradicated the incidence of fraud in Costa Rican elections," writes Costa Rica scholar Charles Ameringer.[22]

Partial public financing of elections helps reduce the influence of rich supporters.[23] And in 2009, the country passed an electoral code that strengthens public funding of elections. Unlike what's now legal here by virtue of the 2010 Supreme Court decision unleashing corporate spending in elections, Costa Rica's new code permits only real people, not corporate bodies, to spend money in election campaigns and requires full transparency.[24]

Costa Rica celebrates Election Day as a national holiday on May 8 every four years. All schools, businesses, and banks close, and all transportation is free, enabling voters to get to the polls. Every citizen over eighteen is able to vote. While voting is mandatory, no penalties are imposed on those who don't—in recent years, about a third of those eligible. One blogger captured the spirit of Election Day this way: "Private cars are expected to stop when someone indicates they want a ride to a polling booth. . . . In practice most autos will stop for anyone who waves and asks for a ride. . . . After all, this is a fiesta!"[25]

One reason for the country's comparatively stable and fair democracy is its economic journey. In the sixteenth century, when, like much of Latin America, Costa Rica was colonized by the Spanish, the countryside didn't end up in big estates. Much of Costa Rica's "agricultural production was carried out by thousands of independent farmers who owned the land on which they cultivated coffee," writes Helmuth. So when the global coffee market boomed, they—not just the big plantations—benefitted.[26] In 2000, after steady decline over thirty years, Costa Rica enjoyed the lowest incidence of rural poverty among seventeen Latin American countries.[27]

Thus, it doesn't surprise me that a country like Costa Rica ranks near the top worldwide in measures of environmental progress. In fact, the big picture view reveals a strong correlation between democratically accountable governments and environmental sanity. Iceland, Switzerland, Costa Rica, and Sweden rank as the top four "green" countries, according to the Environmental Performance Index. All have relatively vibrant political cultures and, with the exception of Switzerland, enjoy high voter turnout.[28]

## AN ECOLOGY OF DEMOCRACY

With courageous citizens in the Middle East, China, Myanmar, and beyond risking all for democracy, let's not give up on ours.

Instead let's acknowledge what's more and more obvious: that our idea of democracy—elections plus a market—is just *way* too weak for the job. Then we can get on with the work of bringing to life an emergent, more powerful understanding of democracy that does work because it's creating a context that reflects what is now clear about human nature: our capacity for cruelty when power is too concentrated, secrecy prevails, and scapegoating ensues, as well as—under the opposite conditions—our capacity for fairness and cooperation as we leave behind the status of whiners and blamers and share in power as doers and creators.

I call the latter "Living Democracy." It is not simply a structure of government, so it can't be achieved merely by ousting the remaining forty or so of the world's autocrats. A constitutional form of government is only a beginning. Living Democracy is a way of life, a daily practice. It is not a set system, finished and done, but a set of system values—with inclusion, fairness, and mutual accountability among them—that infuse all dimensions of our common life.

Such an ecology of democracy sheds the useless, mechanical dichotomy in which much domestic political thinking has been stuck: a liberal tradition that sees the state as provider versus a conservative stance that believes we're best left alone to pull for ourselves. One side focuses on the state, the other, on the market. An eco-mind can't think this way: It sees connectedness in which solutions arise in relationships of mutuality—where all participants are both givers and receivers, and all are held accountable for ensuring rights and fulfilling responsibilities.

In February 2011, two very different thoughts expressed by Egyptians, still in the heat of their eighteen-day revolution, captured for me something of this contrast between the received and the emerging way of seeing democracy. First, student Khaled Kamel, twenty, a member of the team using Facebook and Twitter, along with disciplined nonviolence, to bring down President Hosni Mubarak, told *Time*, "I don't care who ends up running this country, as long as I have the ability to change them if I don't like them."[29] Despite Kamel's revolutionary fervor, I hear strains of the old and failing notion of democracy—a structure over us in which our role, at best, is to accept or reject. The day after President Mubarak acceded to peaceful demands to leave, CNN's correspondent Fredrick Pleitgen tweeted what he was hearing in the streets: "Can't believe the massive cleanup in Tahrir Sq. People tell me: we have won our country back. Now we will rebuild it."

In the latter is the seed of the idea of common responsibility, the heart of Living Democracy. Can we in the US, one of the world's oldest political democracies, reimagine a Living Democracy embodying a sense of co-responsibility, far beyond simply choosing our rulers? That's the question.

"Government used to be the place in our community where people came together and made civic decisions," remarked John Reuter, a Republican city councilman in Sandpoint, Idaho, speaking to the *New York Times Magazine* in 2009. He continued, "That's what we should do again, and that's what's going to bring us back together: not having government be this force somehow outside of us, that's bearing down on us or annoying us, but as a force that we actually embrace and want and that does what we want."[30]

Hmm . . . he's a Republican, and that could have been me talking.

Reuter helps to lead the Sandpoint Transition Initiative, part of the movement noted earlier of almost four hundred municipalities worldwide, preparing this community of 7,500 for a renewable-energy world.

For me, Reuter is talking about Living Democracy. It is a way of being that assumes that the values of shared power, transparency, and mutual accountability—people stepping up, not just pointing fingers—apply not just in the political arena but in all realms of community life, including education, criminal justice, human services, and the economy.

Of these, the most difficult arena of Living Democracy for many of us to envision is economic life, so I've touched on several aspects in this book—from the fast spread of democratic cooperatives, to the "Local First" movement building loyalty to community-connected businesses, highlighted in Thought 1, to rules society enforces to prevent monopolies, to the efforts of labor to protect worker dignity.

What's also not to be overlooked with an eco-mind is the everyday economic power we exert at a point of influence I call "power shopping"—purposefully using our spending to signal businesses to produce more fairly and sustainably. Even in a difficult economy, nearly 70 percent of Americans think it's "important to purchase products with social and environmental benefits," according to an industry survey.[31]

But the UK is the hands-down champ here: Expenditures on what the Brits call "ethical goods and services," including investments, has leapt nearly threefold in a decade. Sales of fair-trade goods—guaranteeing fair returns to family-scale producers and co-ops—are up a phenomenal thirtyfold. Half of all Brits now say they've made at least one ethically based selection in the last year—twice the share a decade ago.[32]

Beyond our solo power of the purse, how about a fervent mob applying power shopping?

In early 2008, Brent Schulkin, twenty-eight, approached two dozen San Francisco convenience stores with this intriguing offer: If you commit to a big eco-friendly business investment—like greening your lighting system or heating—I will guarantee you that a swarm of new customers will descend on your store in a single day. Brent then sealed a deal with the business making the biggest promise. On the chosen day, he used social media to rally hundreds of customers to that store, *tripling its revenue.* Thus, the world's first "carrotmob" was born, and since, Brent's "buy-cott" strategy has spread to some twenty countries.[33]

In the workplace, Living Democracy means empowering workers' voices. And Germany's approach suggests that workplace democracy brings resiliency, as the country's economy seemed to weather the Great Recession better than most. There, clerks and other low-level workers have a voice in management through "work councils"—deciding on store hours, who takes which shift, layoffs, and more, notes attorney Thomas Geoghegan, writing in *Harper's*. Larger firms, mainly those with more than 2,000 employees, are governed by "codetermined boards," to which employees elect half of the members.[34] Also, 62 percent of Germany's workforce is covered by collective bargaining contracts with on-the-job protections. So, compared to most countries, German workers have a voice. This is one likely reason that in the recent downturn, German companies were more likely to divide up the reduced workload so that fewer people suffered unemployment.[35]

And *politics* in a Living Democracy?

It starts with preventing the power of concentrated wealth from distorting political decision making and instead infusing the voices of citizens—in other words, creating truly "publicly held government."

Imagine that.

## WHO PAYS THE PIPER

To get there, we have to face the beast.

And that means we stop kidding ourselves. We stop imagining that any of the mighty challenges we face today can be met without a public decision-making tool accountable to the public interest. We reclaim government from private interests.

Counting only registered lobbyists, roughly two dozen roam Washington's corridors for every single representative you and I have sent there to represent us.[36] Among them are lobbyists for the fossil fuel and nuclear industries, including oil, gas, coal, and electricity. In 2010 spending by these industries on lobbying, $338 million, left even the pharmaceutical industry in the dust.[37] One example is Koch Industries, an oil and chemical conglomerate.[38] It's outspent even Exxon in recent years, as noted in the opening chapter, to sow seeds of public doubt about climate science and to stall legislation addressing climate change and pollution.[39]

Privately held government means that citizens' voices are drowned out by a flood of private-interest money. In 2007, for example, about three-quarters of Americans responding to a BBC poll "favored higher [energy] taxes if the revenues will be used to increase energy efficiency and develop alternative energy sources."[40] Yet, our government hasn't even given the approach serious attention. Instead, continuing private industry power within the public sphere is captured in this contrast: over 4,600 off-shore oil and gas leases approved since 2005 compared to just one off-shore wind project.[41]

Most Americans get it. We grasp how private power is robbing us of democracy—77 percent believe large campaign contributions block government from dealing with our biggest challenges, from health care to climate change.[42]

But we can't give up, and we don't need to.

True, in January 2010 the Supreme Court's "Citizens United" ruling overturned a century of court opinion to unleash corporate and union campaign spending. It stunned many Americans. Eight in ten of us opposed the decision, whose effect is now easy to spot: The next election, 2012, became the most expensive in US history, with overall political ad spending doubling compared to 2008. What made the difference was spending by outside organizations, like Super PACS, which shot up 130 percent.[43] And the source of over 40 percent of such money was kept secret. An anathema to democracy.[44]

But there are still pathways to democracy open to us.

Over the long haul, democracy will depend on courts that understand what average Americans seem to: that unleashing big money isn't "freeing speech" but drowning out speech—that of regular citizens—and, just as critically, democracy will depend on courts that see big money's role in elections as narrowing the range of voices citizens need to hear to make wise decisions.

> Democracy becomes perhaps *the* intrinsic social value, for without it neither justice, nor peace, nor ecological health is possible.

Through its nonrelational frame, the current Court majority understands its role as defending disconnected individuals' right to do as they please—in this case, corporations (deemed "persons") spending at will to further their interests. By contrast, an eco-mind, grasping our inherent connectedness, can perceive democracy as a web of relationships that serves life only if all participants can engage in creating rules enabling each of us to influence the whole. Appreciating the power of context to shape all entities—the whole.

Appreciating the power of context to shape all entities, including us, an eco-mind sees the need to ensure an ongoing dispersion of power so that no one of us can warp the social context in our interest alone. In this view, democracy is no longer simply a means to another end. It becomes *the* intrinsic social value, for without democracy, neither justice, nor peace, nor ecological health is possible.

Given the Supreme Court's ruling, some now believe our only option is to go straight to a constitutional amendment clarifying that "corporations are not persons" and "money is not speech." It requires passage by two-thirds of both houses of Congress and ratification by three-fourths of state legislatures.

*Bottom line?* Even if Congress acted, any thirteen states could kill it.

So I believe there is no "straight" path to real democracy. To even dream of an amendment, we need a big, bold stepwise path. As a community organizer, I learned that making big change means setting sights on believable victories, celebrating gains, then moving ahead. Take the DISCLOSE Act to make corporate political contributions transparent—a big deterrent since most corporations probably don't want to risk alienating their customers. In 2010 it failed by just one vote in the Senate.[45] That we can change! Three-fourths of Americans believe it's "urgent" for Congress to act.[46]

So here's what I'm excited about: becoming part of a national, bipartisan movement uniting us across all "issues." Step by step, we replace the power of money with the power of citizens via an invigorating, unifying, fun movement that's bold, yes, but grounded in staged victories. Great initiatives are already underway. For my take on how they are energizing a movement, see "smallplanet.org/livingdemocracy."

I love, for example, the chutzpah of Represent.Us. The campaign asks citizens first, and then politicians, to co-sponsor a truly comprehensive law, the American Anti-Corruption Act. It enables candidates to run in part by voter-power, because every voter gets a $100 voucher to support a candidate or party. It also seriously limits lobbyists; mandates disclosure of all political money; strictly limits political action committees; shuts the "revolving door" in which legislators morph into lobbyists—and more.[47]

But we also need real-life examples of how democracy can work, freed from corporate control.

So now . . . a story.

## IF DEB CAN DO IT

My hero Deb Simpson of Maine embodies what is called "voluntary public financing" of elections, that's worked for over a decade in her state. [48] In 2000, Deb, a college student and single mom, was working as a waitress in a diner in Auburn, Maine. Friends saw leadership qualities in her and suggested she run for office. "What?" she asked. "I don't have big bucks or a big name." But her friends explained, "We have Clean Elections in Maine. All you need is five dollars each from fifty people to get on the ballot and access public funds."

"Oh," Deb thought. "I can do that. I'm a waitress!" So she ran, she won, and she was re-elected three times to the Maine House of Representatives, where she co-chaired the joint Judiciary Committee. In 2008, Deb won a Maine Senate seat. "Because I survived domestic abuse myself, I've loved being in a position to strengthen laws protecting women," she told me. Deb also served on the joint Natural Resources Committee.

But in 2010, out-of-state money swamped Maine's Clean Elections. Deb lost. Now, her service *and* the manner of her defeat fuel my passion for reform because there are many more Deb Simpsons among us.

## WHAT ACCOUNTABLE GOVERNMENT CAN DO

Mainers used to throw away every year an estimated 100,000 computers and televisions, each with toxic materials like deadly mercury and lead inside. Many of them were incinerated, producing harmful pollution in this breathtakingly beautiful state. That was before early 2006, however, when Maine's first-in-the-US "producer-responsibility" law went into effect. It ensures that producers of certain electronic products and by-products contribute to the cost of recycling; manufacturers must take responsibility for the items they've made even after they are no longer useable.

You won't be surprised when I tell you that as soon as the law was introduced, industry lobbyists mounted the biggest effort to defeat a law that Mainers had ever seen. "All the big computer makers—IBM, Panasonic, Mitsubishi—put tens of millions into TV and print ads opposing the bill. Apple, despite its progressive image, was the worst," Pete Didisheim told me. He's advocacy director for the Natural Resources Council of Maine, which led the campaign.

E-waste is the fastest-growing waste stream in the US. But in the first three years after its passage, Maine's law had kept over 1.5 million pounds of lead (that's more than 1 pound of lead for each resident)—along with mercury, cadmium, and other toxic compounds—from despoiling Maine's ecosystem. The law resulted in the safe recycling of over 11 million pounds of toxic electronic waste. It is also saving Maine towns a lot of money because manufacturers now pay for recycling.[49] Most of the lawmakers voting for this innovative law were able to respond to this commonsense solution, initiated by their constituents, in part because they'd run "clean," i.e. without corporate money. And the mighty ripples? Twenty-five states and New York City have e-waste laws modeled after Maine's—or even going beyond it. By 2012, they covered almost two-thirds of all Americans.[50]

Clearly, common sense can triumph when decision making isn't beholden to special interests. But if you still doubt the US can ever get there, recall what we've accomplished before. In 1976, not big, private bucks but public financing—what made public service possible for Deb Simpson—covered 58 percent of presidential election spending.[51] Today, voluntary public financing is in place in three states—Arizona, Maine, and Connecticut—for all statewide offices; and sixteen states offer some form of public funds to qualifying political candidates for some offices.[52] The Fair Elections Now Act would take this approach to federal elections. The 2012 Grassroots Democracy Act uses a different tack, including a $25–$50 tax credit for voters to back candidates, along with a public multiple-match for candidates refusing PAC money.[53] Represent.Us, above, aims for comprehensive reform.

So whatever our personal passion, we can apply some of that great energy to the root: *getting money out, and our voices in.* If you're tempted to say it's impossible that Congress would vote for a policy that might disadvantage its sitting members—think about what Brazilians pulled off. In Brazil, citizens can propose a law *if* they get 1 percent of voters, not online but real signatures from at least five states, to approve. "That's a lot of standing on street corners," Chico Whitaker, cofounder of the World Social Forum, told me. "Starting in 1997, we gathered a million signers for a law to stop electoral corruption, buying votes," Chico explained. "Congress debated it and, with some changes, approved the law in seven weeks. A record! That was September 1999, and it's been in effect during five elections—eliminating about a thousand candidates who tried to buy votes. A big success!"

Ten years later, he explained, citizen movements demanded that anyone indicted for a crime not be allowed in Congress. "This time we needed 1.3 million signatures, and again it took one and a half years to obtain them. A lot of people fought it, but the law Congress approved was better than our proposal," he said. "Another big victory."[54] One signer wrote on the Facebook page of Avaaz.org, supporting the effort: "Today I feel like an actual citizen with political power. —Silvia."

Also recall Brazil's recent, stunning advance against poverty. So for me, Brazilians' grit holds huge lessons. In 1980 citizen movements formed their own political party that by 2002 had elected a former factory worker president, whom they held accountable for big change. And for us? Key is transforming the "issue" of money's control of our public decision making into nothing less than a citizens' movement for the soul of democracy.

So, visit the websites above, weigh in, and spread the word. Never lose sight of the simple truth that solutions to the multiple threats to our planet hinge on our being as tough as the Brazilians.

## DEMOCRACY, LIVE IT OR LOSE IT

Earlier I included hints of a kind of "animal democracy" in which a whole herd takes part in setting a direction. Scientists suggest that this sort of animal behavior may carry with it a real survival advantage.[55] My hunch is that it's certainly true for human societies: The more inclusive the decision-making process, the greater the amount of information that gets weighed and thus, typically, the better the decision. And to tap this potential, groups must encompass real diversity of views, observes James Surowiecki in his widely read *The Wisdom of Crowds*.[56]

To enable this all-important independence of opinion and the likelihood that the resulting diversity of views is heard, let's peer through the lens of ecology that keeps relationships foremost. As we connect with each other in the life of our communities, the question from an eco-mind becomes, How does my action, right now and for the future, generate relationships of empowerment so that more and more voices are heard?

Whether the issue is replacing dirty with clean energy and transportation or growing healthy food or transforming my children's lunchroom, that's the core question for an ecology of democracy. For if we and others feel powerless, we feel voiceless—killing the possibility of a rich diversity of views needed for wise action.

But how?

The "how" for me begins with remembering that democracy is a skill we can learn. Motivation gets charged when I realize that our multiple challenges—avoiding the worst of climate catastrophe and adjusting to now inevitable climate changes, converting to green energy, and moving from massive poverty to social health—will occupy human creativity for the rest of this century and far beyond. Arguably, we'll either solve them all or none of them.

So now is the moment to focus on building the essential assets we'll need for this passage: At the center are people skills—skills of participation and problem solving, what I call the "arts of democracy"—including judgment, negotiation, mediation, valuing diverse perspectives, and creative conflict. At the Small Planet Institute, we offer *Doing Democracy*, a handbook introducing ten of these "arts."[57]

Top-down approaches deprive us of the opportunity to build these essential capacities, so new strategies are emerging around the world to bring democracy to life from village to shop floor.

## New ways of tapping the wisdom of crowds

Here are a few such proven approaches to tapping citizen smarts, enabling everyday people to contribute to solutions. They could be adapted to any community or institution. How different they are, for example, from the so-called town meetings in the summer of 2009, where opponents of the administration's health insurance reform seemed to think democratic debate meant shouting insults.

A lot of us winced at those ugly scenes. Now we can spread stories of face-to-face democracy working.

*The Citizens' Jury—giving farmers voice.* The Citizens' Jury is an approach to collaborative problem solving in which as many as several dozen randomly selected citizens come together over several days to weigh a critical issue and agree on a direction. Pioneered by the Jefferson Center in Minnesota, hundreds of these juries have worked through tough challenges from sewage treatment to climate change.

As a part of what some call "democratizing the governance of food systems," Citizens' Juries are arising in South Asia, West Asia, and the Andean region of Latin America.[58] And in West Africa, where representatives from fifteen organizations—including farmers groups that have long felt pushed out of decision making, along with government officials, academics, and more—designed the jury process. Before the talking began, an independent farmer-led assessment of public agricultural research arrived at recommendations that fed two Citizens' Juries in early 2010. In each, roughly forty jurors from four West African countries weighed diverse evidence and views and came to clear recommendations about their countries' farm policies.

One: Include us! Farmers want a "central role" in choosing the focus of public research. Another: Avoid genetically modified and hybrid seeds (that can't be effectively saved for replanting). Instead, focus on improving local varieties, ecological practices—including mixed cropping and composting—as well as the exchange and use of local seeds. The farmers call this direction "food sovereignty," the opposite of dependency.

The juries' findings—favoring exactly the approach to agriculture that's best for addressing climate change—got a lot of media attention across West Africa.[59]

*The Deliberative Poll—letting Texans have their say on energy policy.* In the Deliberative Poll, developed by Stanford professor James Fishkin, a larger group—typically a hundred or more randomly chosen citizens—deliberates face to face about a public problem it cares care about. Beforehand, participants review briefing materials collaboratively produced by all interested parties so they are accepted as fair. Before discussion begins, an in-depth poll registers participants' views, and after a day or more of talking through the question, they're polled again. Thus, it's easy to spot changes triggered by new information and discussion.

The Deliberative Poll harkens back 2,400 years to Athens's original use of a council of randomly selected citizens, and in this modern form, it has been used dozens of times in widely diverse cultures, including to select candidates in primaries in Athens itself and to weigh policy directions in China.

In Texas, the Deliberative Poll might help to explain why the state surprised everybody—especially this Texas-reared girl—in taking a big, early lead in wind power. Between 1996 and 1998, under Governor George W. Bush, eight utility companies used Deliberative Polls to develop public understanding and to weigh public opinion. Because participants got to learn together and to think through options with peers, they brought informed opinion to the discussion.

Through the process, the utilities learned that Texans already felt pretty favorably toward renewables. But, in a significant shift, by the end of the process they'd ranked efficiency higher than when they began, and the share of those willing to pay for renewables and conservation increased by more than 60 percent.[60] Preference for fossil fuels sank, reports the National Renewable Energy Laboratory.[61]

Apparently, the utility companies listened, for a decade later, the US "soared past Germany as the world's biggest generator of wind power, thanks to a surge in wind-farm investment led by Texas," *Newsweek* reported.[62]

Hmm. I wonder, Are the "deliberators"—those who more than a decade ago sat around tables telling the utilities what they thought—now cheering?

*The 21st Century Town Hall Meeting—giving 3,500 Americans a voice in tough budget choices.* The 21st Century Town Hall Meeting is the brainchild of Washington-based America*Speaks* founder Carolyn Lukensmeyer, who's got an amazing knack for bringing people together to produce results.

In the approach, 500 to 5,000 people participate in discussions, but Lukensmeyer wants to be clear: This is not a "public hearing"—no Q&A, no panels. Rather, the large body is divided into diverse groups of ten to twelve people who engage in roundtable discussions, each assisted by a trained facilitator. Written guides provide background. Each facilitator uses a laptop networked to other laptops in the room, so each table's ideas and conclusions can be transmitted instantly to a theme team. That team synthesizes all of the information and presents the synthesis to everybody on a large screen. On individual keypads, participants then vote on the synthesized recommendations.

Here is a taste of what can happen.

With support across the ideological divide, from the Center for American Progress to the Heritage Foundation, in June 2010 America*Speaks* created gatherings in nineteen cities. For three hours, a cross section of citizens discussed America's economic recovery and the values they think should guide fiscal policies. Participants also voted on options for reducing long-term deficits, chose messages to send to Washington, and committed to local actions. In broad strokes, the vast majority, 60 percent, said that government should do more to strengthen the economy, and more than half thought the burden of reducing the deficit should fall more heavily on those with financial capacity.[63]

Their findings went directly to the bipartisan White House commission studying ways to cut the deficit. But did you ever hear about it? Not likely. Imagine democracy where a thoughtful process like this would receive visibility and credibility on par with any commission a president handpicked.

*The Citizen Assembly—helping British Columbians choose a new electoral system.* Using large groups and multistage processes, Citizen Assemblies in many countries have taken on huge questions as well.

Over ten months in 2004, for example, 161 randomly selected British Columbian voters researched and debated a range of electoral systems. This Citizen Assembly also held fifty public hearings. In the end, participants overwhelmingly chose a proportional system, like that used in Ireland, in which voters rank candidates, and the votes of those whose candidates don't make a certain cut are transferred to their next preference.[64] When this choice of electoral system went before provincial voters, it garnered 57 percent of the vote. But passage requires 60 percent.[65] Advocates still hope this thoughtful process, with results supported by a strong majority, will, with more effort, bear fruit.

*The Study Circle—enabling 1,300 Kansans to choose how to deal with their trash.* In Study Circles, typically eight to twelve people come together with the help of a trained facilitator to explore an issue in multiple sessions and work together to come up with ideas for action. Nearly 2 million Swedes participate each year in Study Circles, subsidized but not controlled by the government.[66] In the US, Study Circles have tackled tough challenges—from racism and racial tensions following the 1995 O. J. Simpson trial in Los Angeles to worrying high school dropout rates in Kansas City, Kansas.

The chief promoter of Study Circles in the US is East Hartford–based Everyday Democracy. Since 1989, it's worked with more than six hundred communities across the country. In the late 1990s, Study Circles helped officials and community groups in Sedgwick County, Kansas, to engage citizens in making big decisions about what to do with trash. The process involved 1,300 citizens in eighty-two face-to-face sessions, setting the agenda and choosing among waste-disposal options. The citizens picked recycling as their first priority, and the county enlisted corporate help to put in place county-wide recycling. Several months later, eight hundred residents convened in Study Circles again to work through recycling specifics, which the county incorporated into its waste management plan.[67] By 2009, household recycling in the county had significantly increased.[68]

*Participatory budgeting—enabling everyday citizens to deliberate together on how to use public funds.* Launched in Porto Alegre, Brazil, participatory budgeting gathers citizens in a series of open assemblies each year to determine the use of a significant portion of a city's annual budget. Now twenty-five years old, the approach has spread to 1,200 cities, from South Africa to the suburbs of Paris.

Effects have been dramatic, as public spending shifts toward poorer communities that earlier, with more centralized and secretive decision making, had less clout.

Recall my earlier discussion of how we humans do better when we know others are watching us. So another benefit of participatory budgeting is a noticeable decline in corruption under the watchful eyes of so many citizens. Visiting a neighborhood near Porto Alegre in 2003, I admired the big, new community center and heard about a new school and clinic. When I asked, "But how can you afford all this?" I was told by smiling locals that less corruption meant more funds for the community. In other words, the new participatory system means greater government efficiency: In 1988, an administrative dollar in Porto Alegre bought three dollars' worth of services; ten years later, it bought seven dollars' worth.[69]

Even champions of participatory budgeting have lamented small turnouts relative to potential engagement. So, Belo Horizonte, Brazil's fourth-largest city, expanded the process

electronically, allowing a forty-two-day period for those participating to weigh in online. Participation jumped from 1.5
percent of the population to 10 percent. The winning options
included renovating the transportation system and creating
an "ecological park."[70]

Which brings us to . . .

*E-democracy—enabling Estonians to shape their nation's
health plan and Brazilians to choose budget priorities.* In 2001
the former Soviet nation of Estonia mandated that public bodies create extensive websites, posting drafts of pending policies and laws for citizens to review. Several years later, the
country's health ministry reported that its health plan had
been amended radically in response to citizen input.

Online connection can spur people to connect off-line too.
In Estonia, population 1.3 million, this potential became real in
2008: When alerted via the My Estonia website, 50,000 volunteers joined together in one day to collect 10,000 metric tons of
"dumped garbage." (Translated proportionally to the size of
the US, that would be over 11 million people turning out!) And
in 2009, 11,000 citizens convened brainstorming sessions
under the slogan "Thinking Together for a Better Estonia."[71]

Brazil's House of Representatives is also an e-democracy
leader. Its website lets users access not only text but also
audio and video related to pending legislation, and via its
"wikilegis" area, citizens can suggest amendments. The site
encourages users to create their own personal profiles, so that
thematic social networks can emerge too.[72]

Imagine these and similar practices for reaching public judgment helping Americans set priorities and guide policy choices in the years ahead.
Beyond imagining, why not insist that anyone running for office commit
to tapping such proven approaches and taking them to scale? That way,
we leave behind the finger-pointing of today's political culture and begin
to assume mutual accountability for solutions, one of the three essentials
of Living Democracy.

But in ticking off these six road-tested, inclusive methods of coming up with better public choices, I don't want to jump over the less-formal, vital, and growing roles citizens are playing right now that rarely make the evening news, citizens like those in the many stories throughout this book—from Maine, where citizens of three small towns were victorious over the giant Nestlé corporation in protecting their communities' water supply, to Chicago, where Growing Home works with diverse citizens to turn a "food desert"—bereft of much except junk food—into a "food destination."

These and millions more of our neighbors are already *doing democracy*. Some of my favorite examples you'll find in the "moving into action" guide at the end of the book, with a note about each initiative. Among the most underappreciated are the roughly 3 million Americans who participate in more than 130 largely faith-based, multi-issue, interdenominational community organizing networks—like the PICO National Network, which is now in 150 municipalities from coast to coast. These networks are tackling major public concerns, from education reform to health care. Greater Boston Interfaith Organization, an affiliate of another network in this movement, is celebrated for its leadership in helping Massachusetts become the first US state with virtually universal health-care coverage.[73]

And doing democracy includes the almost twenty gutsy citizen groups, coordinated over two decades by the Citizens Coal Council, in states most hurt by coal. They are taking on the industry, and we should all be grateful: Recall the 2011 study estimating that coal's impact costs Americans about one-third of a trillion dollars a year. In 2000, one of the council's affiliates, Save Our Cumberland Mountains, succeeded in convincing the Office of Surface Mining to make 61,000 acres, almost three-quarters, of Tennessee's Fall Creek Falls State Park watershed off-limits to mining. It's the biggest area ever protected in this way.

None of these citizens appears to feel "it's too late."

## DEMOCRACY AND DIGNITY

In early 2011, democratic uprisings in the Middle East caught a lot of Americans by surprise. Is it that some of us imagine "freedom" and related values to be especially American? If so, we could be missing something big—a key aspect of human nature itself.

Decades ago, in an international survey ranking 42 values people hold dear, dignity ranked high but still beneath freedom and justice.[74] Perhaps, however, sensibilities are evolving. Since the beginning of the Arab Spring, cries for "dignity" have intensified. For me, dignity is both foundational and the most democratic of sensibilities. I can imagine a dictator, after all, providing everyone food and other essentials but the populace feeling that dignity has still been denied. Or, I can imagine being left alone and therefore literally being "free"; but with no power to meet my needs, I could lack dignity.

Seeming to affirm these findings, the twenty-year-old Egyptian student I quoted earlier, Khaled Kamel, spoke of his motivation rooted in the indignities he experienced under President Mubarak's regime: "The police treated us like we were a different kind of human," he said. Once, after falling off a train, he recalls that a policeman, "instead of helping me, he hit me because I was lying there on the platform, which you're not supposed to do."[75]

His feelings have been expressed by many—from Wal-Mart workers on Black Friday in 2012 demanding not just fair compensation but basic human "respect," to indigenous people in the Ecuadorian Amazon demanding compensation from Chevron for polluting their sacred home, the forest.

For me, dignity has always carried within it the assumption of "voice," of being respected enough to have one's voice count. Dignity is about feeling valued, not as a means to another's end but as a being of inherent worth with something to offer. Dignity is therefore an eco-concept. It lacks meaning except in relation to others.

We cannot know how events in the Middle East will unfold, but we do know that in early 2011 in eighteen days, tens of thousands of Egyptians achieved not just the end of something odious but the beginning of new dignity.[76]

This chapter began declaring that "it's not too late for life" and moved to our imagining a Living Democracy, a process of engagement in which we are together dispersing and co-creating power as we share in responsibility for our future so that *all* can experience dignity. That's what each of the specific, far-flung examples above is all about: the dignity of feeling counted as part of our era's historic search for solutions.

Thought Trap 7—"It's too late!"—is of course factually correct. It is too late to avoid major losses that actions taken even a few decades ago could have avoided. But it is not too late to take in all we're learning and to trust—to trust that we now know enough about ourselves to believe that a deepening of democracy is within us. To trust that we can see the environmental crisis as in fact a crisis of democracy itself and that, with elegant symmetry, ecology—its truths of connectedness, dispersion of power, and continuous change—can show us the way.

Along this path we may discover that democracy—real, Living Democracy—is not just a society's best strategy for making decisions. It is also a different way of understanding ourselves, a way more fully in tune with the rich complexity of our own nature.

## THOUGHT LEAP 7

It's not too late for life. And its flourishing depends on democracy attuned to all we now know of our nature and to what history teaches: that centralized power cannot lead to the sustained good of the whole. Plus, today's crises are so interrelated, vast in scope, and complex that solutions depend on tapping the direct experience, common sense, and passion of people most affected. And that's billions of us! With the buy-in that comes with having a real voice, most of us want to contribute to solutions that turn our planet toward life. So the "mother of all issues" is removing the power of concentrated wealth from public decision making and infusing citizens' voices instead. It's beginning in surprising places.

# An Invitation—Thinking Like an Ecosystem

*I can't understand why people are frightened of new ideas.*
*I'm frightened of the old ones.*

—John Cage, twentieth-century composer[1]

## The work of hope

HOPE IS NOT WISHFUL THINKING. IT'S NOT A TEMPERAMENT WE'RE BORN with. It is a stance toward life that we can choose . . . or not. The real question for me, though, is whether my hope is effective, whether it produces results or is just where I hide to ease my own pain.

What I strive for I call "honest hope." And it takes work, but it is good work. It is work I love. I began this book suggesting that it starts with getting our thinking straight. Since we create the world according to ideas we hold, we have to ask ourselves whether the ideas we inherit and absorb through our cultures serve us. We can only have honest, effective hope if the frame through which we see is an accurate representation of how the world works.

The good news is that we face this historic challenge just as our understanding of life's rich complexity, and human nature itself, is expanding exponentially. I am pretty sure, for example, that I'd never even heard the word "ecology" until I was in my twenties. And that was only because I was fortunate enough to marry one of our country's most brilliant ecological thinkers, the late Marc Lappé. Now we are realizing that ecology is not merely a particular field of science; it is a new way of understanding life that frees us from the failing mechanical worldview's assumptions of separateness and scarcity.

So here, in this final chapter, is an invitation to explore what it means to think like an ecosystem. Since ecology is all about interconnection and unending change, creating patterns of causation that shape every organism and phenomenon, "thinking like an ecosystem" for me means living in the perpetual "why." It's keeping alive the two-year-old mind that accepts nothing simply as "the way it is" but craves to know how something came to be. It's understanding that all organisms emerge with specific potential, including the human organism, but its expression is enormously shaped by context.

So, if we want life to thrive, we keep foremost the question, What conditions enhance life? And, more specifically, what specific conditions bring out the best in our species? My hypothesis is that three conditions—the wide and fluid dispersion of power, transparency, and an assumption of mutual accountability—are at least a good part of the answer. An eco-mind is also able to see that our own species' thriving, through our *consciously creating the essential context for that thriving*, determines the well-being, even the continuation, of other species and whether key dimensions of our wider ecology remain conducive to life.

Shifting from the mechanical assumption of separateness and seeing our societies as ecosystems, we get curious about how aspects interact. And, writes Oxford historian Theodore Zeldin, "It is only curiosity that knows no boundaries which can be effective against fear."[2]

Now, there's motivation.

## OUR CULTURE'S FOOD CRISIS
### THROUGH A RELATIONAL LENS

Using our eco-minds, we soon realize that in our complex human ecology, many of the most important causal interactions may not immediately meet the eye—just as they don't in the wider ecosystem: When you or I look at a forest, for example, we see distinct trees. We don't see that beneath the forest floor trees intermingle for mutual support, sometimes through their roots, sometimes through "mats of cooperating fungi," explains the late sustainability genius Donella H. Meadows.[3] Mycelia, the underground part of fungi, can spread "cellular mats across thousands of acres."[4]

The implication? Cutting one tree is never about just cutting one tree. Every act has multiple effects.

So, let's try taking one huge, specific problem—our country's food crisis—and see what happens if we use our eco-mind to look for patterns of connection.

First, diagnosing the problem using a morality frame or a mechanical "more-or-less" frame doesn't get us far. John Naish, for example, writing in *Resurgence* magazine, tells us that obesity and diabetes flow from "over-striving and over-consuming."[5] He sees through the "too-much" lens.

Through this more-or-less lens, the answer to our obesity epidemic—costing the US and Canada roughly $300 billion each year and contributing almost as much to the US disease burden as does tobacco use—could be to eat fewer or burn more calories.[6] But Americans could do both and still be made ill by our food. To get anywhere, we have to let our curiosity lead us, exploring the interacting incentives and relationships that combine to create ever more unhealthy food "products," more unhealthy eating, and more waste.

In *Diet for a Small Planet*, my favorite chapter—"Who Asked for Froot Loops?"—is one attempt. In it, I ask readers to pretend they head a food conglomerate, so they have one rule to follow, one job to do: to bring the highest return to their shareholders and executives. Every step degrading our food flows from that rule: the more processed the food, the less nutrition but the more profit it offers, and the more fat, salt, and sugar it contains. A consequence is that now 40 percent of the calories US children eat are nutritionally empty.[7]

But this understanding only begs the next question: Why are we humans so apt to overeat foods that aren't good for us?

Recent experiments at Scripps Research Institute shed light on this question and on why the road to junk food has been so lucrative for its manufacturers. It turns out that junk food—high in fat and dense in calories—stimulates the brain's pleasure centers and changes its chemistry, eliciting in rats addictive behavior similar to what humans experience with cocaine. Rats fed Ho Hos, sausage, pound cake, bacon, cheesecake, and the like by researchers Paul J. Kenny and Paul M. Johnson soon developed "compulsive eating habits" and became obese.

Even scarier?

When these rats received an electric shock upon eating the high-fat food, those that had not already been exposed for some time to the junk food stopped eating, but rats used to the junk food weren't daunted: They just kept going for it, shocks be damned. Then, when denied junk food and left with healthy food as their only option, obese rats simply stopped eating. "They starve themselves for two weeks afterward," neuroscientist Kenny reported. "Their dietary preferences are dramatically shifted."[8]

Long before these alarming findings, the food industry grasped their essence. One particularly candid agribusiness insider explained to former Food and Drug Administration chief David Kessler that his industry is "the manipulator of the consumers' minds and desires."[9]

Next, we can begin to perceive what ecologists call "positive feedback loops" that accelerate a trend.

The logic of the system becomes self-reinforcing: As the success of advertising hooking into our brains' circuitry concentrates the wealth of a few large food firms, companies' advertising budgets swell, giving them even more resources with which to entice us to buy what brings them still greater profit—processed foods with the least costly ingredients. Their advertising power plays on another consequence of a culture concentrating private wealth while denigrating the public sphere, including public education. Some cash-strapped public schools have even resorted to selling ad space for sugary cereal in the lunchroom.[10] One-rule economics—highest return to existing wealth—also feeds a society's rapid divide. The numbers of both the very poor and the superwealthy expand, with big diet consequences. For one, it makes little profit sense to build supermarkets where most people are poor, so convenience stores, packed with fattening, processed foods, are about all there is within reach of increasingly overweight poor populations. Today, among the ten poorest US states, seven are among the ten that also suffer the highest rates of obesity.[11]

A one-rule economy also lengthens the workday and workweek, with big consequences for our bodies' well-being, since Americans have to put in longer hours to compensate for lower real wages. Between the late 1960s and the late 1980s, the average American worker added "the equivalent of an extra month" of work each year, says professor Juliet Schor.[12] Sleep is one thing sure to go—we're averaging one to two hours less per night than we did twenty to thirty years ago. And scientists have discovered that sleep loss itself causes people to eat more—and to eat more unhealthfully. One study found that after just two nights of curtailed sleep, subjects were more drawn to candy, cookies, and cake.[13]

Highly concentrated wealth also brings with it economic insecurity that affects how we eat. A twelve-year study, for example, discovered that when their income drops, American men gain on average 5.5 pounds.[14]

Plus, a one-rule economy creates effective monopolies among traders and processors; they then put on so much lobbying muscle that they can reap huge tax subsidies, which help keep the price of their key raw materials, corn and soy, often even below the cost of production. This concentrated industry also deprives farmers of bargaining power, driving them to produce ever-greater volumes in ever-bigger farming operations, just to stay afloat. Then, in our world where a billion people are too poor to make market demand for the food they need, grain and soy flood a stunted market.

In such a world, it's made perfect "economic sense" for US agribusiness to use most of the corn and soy for livestock feed and processed food—giving us fatty meat and ubiquitous high-fructose corn syrup, which show up not just in soda but just about everywhere, even in a Starbucks egg-salad sandwich I bought not long ago.[15]

My point is that only by thinking in systems would we possibly link America's obesity epidemic with tax subsidies—$4 billion a year for corn alone—benefitting the largest industrial food operations or with business lobbying that's often defeated proposed increases in the minimum wage, so that in recent decades it has lost a quarter of its value, which means longer hours for the same pay.[16]

All with major health consequences.

So answers don't flow from thinking "more or less" but thinking "relationships"—asking how we can reclaim democratic decision making to shape smarter rules, rules that align the food corporation's and the farmer's incentives with our well-being. Thoughts 5 and 7 suggest we can.

## Reconnecting,
### but this time with intention

For most alive today, thinking like an ecosystem is, understandably, a challenge. In many preindustrial societies, as well as in many remaining indigenous cultures, it was different. Ecological wisdom lived in, and was passed on through, everyday activities, myth, and ritual. Earlier humans didn't have to learn to "think like an ecosystem" because they'd not lost their sense of living within the ecosystem.

For over 1,000 years in Bali, dense networks of handmade irrigation canals, called *subak*, have made rice growing possible on steep, terraced terrain. Today, rice planting and harvesting, as well as the timing, volume, and sharing of water flowing through the *subak*, are embedded in religion and ritual. Century after century it's worked without degrading the island's breathtaking mountain ecology. Yet, most Balinese probably couldn't tell you why the *subak* were originally constructed as they are or, in scientific terms, how the system aligns with nature.

In a similar vein, the late Nobel Peace Prize laureate Wangari Maathai, who so inspired my daughter and me on our journey to write *Hope's Edge*, loved to tell a story about embedded wisdom. As a child in Kenya's Rift Valley, she played for long hours in a stream near her village. Collecting firewood was her daily chore, but her parents warned her never to disturb one particular tree, the fig—it was sacred.

Years later, Wangari returned to her village only to learn that export crops had replaced trees, including the sacred fig. She observed eroded soils and saw that her beloved stream had dried up. By then a biologist, Wangari was able to appreciate the wisdom in her mother's warning, for it had been the fig tree's deep roots that had helped the soil resist wind erosion and allowed rain to seep into the earth, feeding the stream she'd played in.

Today, what has long served us as unconscious ecological wisdom—developed culturally over eons—we can make conscious. We can develop it with intention, and we can do it quickly.

It is happening.

Take just one breakthrough, a new rice-growing approach called SRI—for System of Rice Intensification. With it, small farmers are using fewer chemicals and less water while producing more food. And all this with as little as *one-tenth* the seed, greatly benefiting the poorest farmers. In about a decade, at least a million rice farmers are using the approach—and are continuing to develop it among themselves.[17]

SRI grew neither from a fancy corporate lab nor ancient tradition—it actually overturns many long-held, unquestioned rice-growing practices. Most agricultural scientists have been "dismissive" of the approach, observes Cornell professor Norman Uphoff, who admits that he, too—now a champion of SRI—began as a skeptic of what seemed to him "fantastical."[18]

SRI grew out of the curiosity and drive of the late French priest Henri de Laulanié in Madagascar. His "keen observation of deviant practice and continued experimentation" led to this "radical departure" in the way rice is grown, writes Indian scholar C. Shambu Prasad. De Laulanié just wanted to help peasant farmers grow food without having to depend on purchased inputs they couldn't afford. SRI isn't a fixed technology but a set of synergistic principles, Prasad stresses, and several of its principles—the wide spacing of seedlings and never allowing continuous flooding of fields—overturned centuries of custom.[19]

It's worth noting that the French priest was a real outsider, unfettered by the "it's-always-been-done-this-way" blinders we all carry. So maybe his curiosity had freer rein.

Thinking like an ecosystem shifts our vision from assuming "trade-offs" to searching for synergies. In what's called "agroforestry," described in Thought 3 as one key in correcting the carbon imbalance at the root of climate disruption, farms and forests are no longer competitors. Rather than continuing to raze forests to grow row crops or graze cattle, some small-scale farmers in Central America, for example, are embracing agroforestry to make their farms more productive and resilient.

In the rugged highlands of northwest Guatemala, for example, Mayan farmers in the village of Guaisna intersperse Andean alder trees right along with their traditional corn crop. The Cambridge-based EcoLogic Development Fund (with which our Small Planet Institute shares offices) worked with the farmers and helped them see the advantage of planting the trees horizontally along the natural contours of their steep mountain slopes. The fast-growing alders' presence has been widely felt: They help prevent erosion—avoiding deadly mud slides—replenish soil nitrogen, and their leaf litter builds the soil's moisture and proper acidity. Plus, their foliage attracts birds, whose voracious appetites reduce the need for pesticides. Over time, farmers' yields increased, while branches cut from the trees became a fuelwood source that now saves villagers long hours searching for fuel. Soon, the shade of the alders and the more fertile soils allowed farmers to add a cash crop, coffee trees, right alongside the alders. Now, with more food and more income, farmers feel less pressure to cut down the forest for new fields.[20]

It all works as a carefully integrated system.

And it wasn't long before farmers in neighboring villages in the Sierra de los Cuchumatanes in Huehuetenango saw life changing in Guaisna and began applying the same approach, one in which a mosaic of forests and sustainable farming is improving lives and land while helping rebalance the earth's carbon cycle.

## IT'S THE CONTEXT, STUPID!

Thinking like an ecosystem means seeing everything in context, or at least giving it our best shot. By this I mean that, with an eco-mind, we realize that what's "good" in one context might bring disaster in other.

I first think jatropha.

Never heard of it? Jatropha is a small tree whose nonedible but oil-rich seeds can be turned into a clean fuel. In parts of rural Africa and Asia, this oil liberates small farmers from hours of daily wood gathering and continuing forest loss. It grows well in poor soils with little rainfall and can be interspersed with other crops, helping to prevent erosion. The tree's smell repels hungry animals, protecting nearby crops. And a poor farmer selling jatropha oil in the West African country of Mali, for example, can double his or her income in the first year of planting jatropha, while not significantly decreasing yields of other crops in the same field.[21]

Jatropha needs no pesticides and no fertilizer beyond the residue returned to the soil after oil is pressed from the plant's nuts. Compare those gains with other biofuel plants like corn or sugar cane, which actually displace crops that could feed people directly and use huge amounts of water, fertilizer, and pesticides.

So what's not to like? Poor farmers, big winners—and the environment benefits too.

Now, place the very same plant in another context.

Several years ago the government of India began backing the spread of large jatropha plantations, with the ambitious goal of producing enough biofuel from its seeds to significantly reduce dependency on imported oil. Here, empirical results from the southern state of Tamil Nadu reveal that jatropha cultivation is not pro-poor at all, say scholars writing in the *Journal of Peasant Studies*. "*Jatropha* cultivation favors resource-rich farmers," they write.[22] Instead of aiding the growth of food crops, as in the Mali story, jatropha plantations in India have replaced food crops and helped push poor farmers off the land.[23]

This contrast in outcome reflects the web of relationships in which the plant grows.

Once seeing through a contextual eco-lens, we also realize that what might be perceived as a single change in a community—whether animal, vegetable, or mineral—can create endless ripples. Hearing, for example, the word "organic," a lot of us see green—maybe a curly kale in a lush field. For most people, it's a package of farming and eating without pesticides. But as we learn to think like ecosystems, the word "organic" can come to evoke vastly wider associations. A recent UN study, *Organic Agriculture and Food Security in Africa*, beautifully teases out a few organic ripples that might surprise.[24]

African farmers who cultivate beneficial insects to control pests, the report finds, develop a lot more knowledge and skills than they did by just spraying pesticide. Where farmers are building on indigenous knowledge, they also experiment more to solve problems instead of merely relying on what a corporate input supplier tells them.

Imagine the greater self-confidence and resiliency in facing climate challenges.

Organic farming also leads to improved health, the report notes, including less malaria in rice-fish zones. Plus, the improved nutritional value of organic produce, along with a greater variety of foods, strengthens people's immune systems, particularly vital to HIV/AIDS sufferers. "Extending the life of a farming parent [with the disease] by several years could mean the difference between life and death for the children left behind," notes the report. Imagine the ripples, where 11 million children in sub-Saharan Africa are already HIV/AIDS orphans.

Organic farming "can undoubtedly reduce poverty" because of increased production selling at higher prices. And because some of the additional income from greater food production goes to paying school fees, "education of the wider community" increases, notes the study.[25]

And then there's the female ripple. In many communities using imported seeds and chemicals, women on their own couldn't get access to these inputs or credit to buy them. (In Africa women receive less than 10 percent of the credit going to small farmers.) But once taking up organic practices, and thus freed from dependency on credit, women gained more equal footing with men. Their output could then grow, providing surplus to sell in the market and helping the whole family.[26]

This report also tells us that, because organic practices can produce more food, hunger is displacing fewer people from their communities. (A separate University of Michigan study found that if the entire world were to move toward agroecological approaches, food output could increase substantially.)[27] Some people are even migrating back to their organic farming villages, as more vibrant local economies offer more jobs.

Finally, climate: In my daughter Anna Lappé's *Diet for a Hot Planet*, we learn about why the food and agricultural system contributes about a third of the gases heating our planet. So, as explored in Thought 3, this move away from extractive, chemical farming means facing down climate chaos. Now, there's a really welcome ripple.

*Clearly, one change—organic farming—is not* one *change at all.*

Even M. S. Swaminathan, celebrated Indian champion of the 1960s Green Revolution—or what I call "dependency agriculture" because it made farmers dependent on corporate-controlled chemicals and seeds—now recommends the direction this report from Africa highlights: toward technologies "rooted in the principles of ecology, economics, gender and social equity, employment generation, and energy conservation." He calls for research "based on an entire farming system."[28]

For me, Swaminathan's shift of perspective is yet more striking evidence that it's possible for any of us to rethink even long-held assumptions.

## READY FOR SURPRISES

As I learn about these breakthroughs, it helps me to keep in mind that ecological time doesn't move in neat, predictable increments. We can be amazed by its glacial pace, just as we can be equally stunned by a sudden cascade—when forces reach a tipping point, bringing on change at gale-force speed. No one, I think I can safely say, predicted that bringing down the Berlin Wall would follow from actions that peaked in little more than a month. Or that Egypt's Hosni Mubarak's three-decade regime would fall after just eighteen days of nonviolent protest. Even a decade ago, very few people, even scientists closely watching our climate's heating, predicted the current rate of polar ice melt.

Throughout, I've tried to strengthen confidence that we can do the deeper work of re-aligning with nature's laws and not allow ourselves to be tempted by the false promise of quick fixes. To keep my own chin up, it helps to carry around a mental list of big surprises, changes I wouldn't have given much chance of *ever* happening. Then, regularly reviewing my list, I have to admit that, well, I could be wrong to doubt again.

It is humbling. But it's a hopeful humility that's growing in me. Here are a few items on my "surprises" list:

Aware that each year forested land the size of Costa Rica is destroyed, in 2006 the UN's Plant for the Planet campaign set a global goal of planting 1 billion trees a year, starting in 2007. Way too ambitious, I thought, but I wished them the best.

Was I off! By early 2011 the campaign had surpassed *11 billion*—that's 1.6 trees for every person on earth.[29] Skeptics tell me, "Oh, sure, but a lot won't survive." And I say to them, not only has the UN Environment Program taken some loss of trees into account, but, hey, even if a big share of the trees fail, this achievement is extraordinary.

Ethiopia alone can boast one billion of them. Over the twentieth century, tree cover in that country had shrunk from roughly 40 percent to just 3 percent. This drastic loss triggered government action encouraging all citizens to plant trees. With a lot of help from the nation's Boy Scouts, over 1.5 billion indigenous tree seedlings were transplanted to speed the effort.[30]

Ready to add your own tree to the global 11 billion? You can register it at the UN Environment Program's website.[31]

> Why is it impossible? We have got 80 million people. If everybody gets up and works, this is not impossible at all.
>
> —Girma Wolde Giorgis, president of Ethiopia, 2009

And there's Bogotá, Colombia, which for too many Americans only conjures up scary drug wars. But since 2000, in this city of more than 8 million, fancy new buses on dedicated corridors handle 1.5 million trips a day.[32] Passengers board through enclosed stations and swipe cards in turnstiles, so the system feels like an above ground subway. Called TransMilenio, it's enabled Bogotá to get 7,000 small, private buses off the streets and to cut the use of bus fuel, thus its emissions, by roughly 60 percent.[33] TransMilenio's initiating mayor, Enrique Peñalosa, also oversaw the creation of extensive bike lanes, 1,200 new parks, and many other public facilities.

And surprising everyone, these big changes were accompanied by a drastic drop of about 75 percent in the city's murder rate.[34] One explanation is that rising trust in city government, having proven itself capable and responsive, made citizens more willing to cooperate with police in preventing and prosecuting crime. And many more families enjoying the outdoors together—possible because of the new parks and walking and biking corridors—probably helped too. Bogotá's striking success has sparked copycats: So far, thirty-seven nations have sent teams there to explore how the city did it, and five Latin American cities have transformed their own bus systems.[35]

Or how about Chattanooga, Tennessee, certainly nobody's pick to become an environmental standard setter?

In this gritty industrial city—deemed to have the nation's worst air quality in the 1960s—drivers had to turn on their headlights in midday. But from its early commitment to citizen involvement, starting in 1984, the city's turnaround has involved thousands of citizens, sitting together to discuss and set goals for their economic and environmental future. The process, called "visioning," helped cut greenhouse gas emissions by the city's transportation fleet by 29 percent, in part due to free downtown electric buses serving a million riders a year.[36] Next in sight are a zero-emissions eco-industrial park and a grass-roofed convention center.

Now, I'd be the last to question whether Chattanooga can do it.

And if that city's speedy environmental resurrection surprises you, consider Pittsburgh. As its polluting steel economy faltered in the 1970s, Pittsburgh made "green" buildings part of its revitalization plan. By 2007 it was ranked as the tenth-cleanest city in the world.[37]

And don't forget Germany, a global green leader popping up often in this book. In the early 1990s, Germany had virtually no renewable energy, but it's now on track to get 35 percent of its electricity from green sources in ten years.[38]

I also remind myself of the dramatic takeoff of fair trade, in which shoppers' choices help ensure living wages and ecological sustainability. It's emerged in no time—just a couple decades. Keeping in mind how much the Brits must shell out in their ubiquitous pubs, let this measure of fair trade's impact soak in: In 2005, sales in the UK of fair-trade goods—for example, one in four bananas sold—and in what Brits call the "ethical sector" were greater than alcohol and tobacco sales![39]

Or consider the land reform movement in Brazil that I mentioned earlier. Its 1.5 million members have created 2,000 new democratically organized communities, with new farms and businesses, many of them organic. Thirty years ago I would have placed Brazil among the least likely to engage in serious land redistribution. Now it is a world leader.[40]

I was *really* wrong.

And I guess I just have to get used to it. For the eco-mind, the one thing that should never surprise is a surprise. The eco-mind sees life as continuous change, with numerous forces interacting to shape any given moment's unfolding.

How freeing. For me this means that *it is not possible to know what's possible*. We are free to follow our curiosity, digging deeper, taking the next step, never, ever able to predict outcome. Like Tyler Hess, a student I met recently at DePauw University, who, with others, had collected nine hundred signatures in a month calling for a campus ban on bottled water. He didn't know whether the administration would go along. It did.

Tyler didn't know whether his school's ban would trigger other Indiana schools to join in. He didn't know whether US college and university actions would then ignite national reaction against this ecological nightmare. And he still doesn't know. But he does know that without his action, the next and the next would be less likely.

With an eco-mind, we keep moving, but not in random motion. So "following our curiosity" doesn't quite describe it. We *pursue* our curiosity, fervently, persistently, asking, *Why*? In our pursuit of how and why we got into in this mess, causal patterns become ever clearer, and we can weave that understanding into our choices.

We do what we love but with ever more clarity—and *guts.*

## Upping our courage quotient

But how do we get more guts?

For me, the idea of thinking like an ecosystem can stir warm feelings—not necessarily braveheartedness. I imagine myself resting on soft pine needles gazing into a dense forest, my body thanking me by becoming healthier each minute, but I also know that much of the challenge we humans face isn't so warm and fuzzy. Yes, it's now shown, we evolved to respond to nature, to cooperate, to empathize, to need to feel useful, and to yearn for meaning in our lives—all this good stuff.

But we're also—and I might as well be blunt—cowards. Most of us are, anyway.

I hope you are not offended. Please let me explain.

I mean that, yes, it's true, many of us would rush into that burning building to save a stranger. Yet, precisely because we are so social by nature, so densely embedded in community ties, just about the scariest thing for most of us is to *be different.* We evolved in tribes, after all, and there we learned we were totally dependent on the group. Banishment meant death.

But today our tribe is taking us down. The hypertribe, the global culture of intimidating, centralized, corporate dominance driven by a myopic "one-rule" market failing to register real costs, is headed right over Victoria Falls. If we just cling to the canoe, we're done for; if we don't jump overboard or convince the crew to head rapidly to shore, we all die.

But breaking with the pack in this way is terrifying for humans. The fear response is of course supposed to help us survive. So here lies a startling question: *Could we be the only species for which fear works against our survival?*

If so, to make this shift to live within an ecological worldview, we have to reframe not just environmental messages but fear itself. Among all the human traits we need to cultivate, we must place first what I now call "civil courage." I first heard this great term in a lecture by professor Otto Herz in Leipzig, Germany, in May 2009.

Earlier that same day I was enjoying a quick tour of this beautiful city when my young guide excitedly pointed to the church of St. Nicolai. In Leipzig, that's where it began, she told me. That church played a key role in bringing down the Berlin Wall.

I was startled. I had never heard that Leipzig, much less this church, was part of the fall of the Wall, on November 9, 1989, leading to the collapse of the Soviet Union and changing the world forever.

Oh, yes, she went on, this is where people began to gather. By early October, protesters in and near the church had swelled to some 70,000. In other German cities, similar protests began.

Now let me continue the story, not through my guide, who was too young to remember, but through Markus Laegel, a fourteen-year-old writing in 1989 of his and his parents' experience:

> Thousands of people gather. They're here to pray. They're praying for peace. The prayer is set to start at 6 p.m. The idea is to use the Nicolai church, but that's already full and it's still only four o'clock. We go down to the church of St. Thomas. That's full too, but we manage to find a space. Mum, Dad and me. We're a family. Not just us, but everyone. Everyone in EVERY church at that moment.

Markus then quotes a minister's words to describe what is unfolding. "When you hold a candle you need both hands. You have to guard the flame. . . . You can't hold a stone or a club at the same time. . . . The army, fighting patrols and police were drawn in, started conversations and re-treated."

And then he continues:

> Thousands of people . . . lay their candles at the feet of the armed soldiers
> and police. The steps of the Stasi [state security] building—the organiza-
> tion that spied on, abused and sold people out—are now awash with can-
> dles. It looks like a river of peace and light.[41]

Exactly four weeks after Markus wrote this account, the Berlin Wall
fell, and soon thereafter a leader in the collapsed East German regime,
Horst Sindermann, acknowledged, "We had planned everything. We were
prepared for any eventuality. Any except for candles and prayers."[42]

How do you feel reading this account? And how did you feel in early
2011 when the Middle East began to break open as everyday people stood
up against autocrats? What is your explanation for how regular citizens—
most of whom had zipped their lips under totalitarian or autocratic control for decades—could bring down, some peaceably, those among our
era's most ruthless rulers?

I share Markus's story because we each need to answer this question.
First, we need to believe that potential courage—civil courage—is in each
of us. And second, we need a theory of how to tap it. Civil courage is different from physical bravery—although it sometimes calls for exactly that
too. It is the capacity to do what is right regardless, even if we have to
stand alone: For even if ultimately tens of thousands joined in, these extraordinary actions depended on the courage of the *first few* willing to step
up, to be different.

My argument is that humans are plenty *good enough*. But we do need
to work on one thing: more backbone. For when put to the test, most of
us will betray ourselves before we will stray from the group. In famous
experiments by Solomon Asch, most subjects went along with the
group's opinion even when it defied what their own eyes were telling
them.[43] Note that often all it took was one truth-teller, however, to enable
people to be true to themselves. So even in everyday situations, James
Surowiecki underscores in *The Wisdom of Crowds*, "one key to successful

group decisions is getting people to pay much less attention to what everyone else is saying."[44]

Fortunately, we are not destined by instinct to shrink from things that scare us. We can use the power of ideas to trump instinct. For the sake of an idea, we see humans overriding biological impulses all the time. Extreme examples include denying oneself food in a hunger strike for the sake of the principle of justice or the horrifying example of murdering an offspring for the sake of the idea of honor.

So why not put the power of ideas to work toward enabling life on earth?

Fear itself is in part an idea. It doesn't have to hold us hostage, and in *You Have the Power*, coauthor Jeffrey Perkins and I offer seven ways to break free. The first is to recognize that while we often interpret fear as a verdict—telling us that we're in the wrong place, doing the wrong thing— we can redefine fear. We can understand it simply as information and energy that we can use *as we choose.*

If stepping out, speaking out, causes our body's fear sensations to flare, why not reinterpret them as a signal? Perhaps a signal telling us we are doing exactly what we should be doing for our own greatest happiness and the planet's.

That pounding heart? Why not rethink it as "inner applause"?

We can use our social nature to our advantage too. As Asch's experiments confirm, often all it takes is one single ally to enable us to stay true to ourselves. Knowing this, we can choose to seek out those who share our passion, those who encourage us to risk for what we believe in.

In a 2008 University of Virginia study, students with weighted backpacks were put at the foot of a hill and asked to estimate how steep it was. Some subjects stood next to friends; others stood alone. Students with friends next to them estimated the hill to be less steep than did those standing by themselves, and "the longer a friend was known, the less steep the hill appeared."[45]

Buddies matter.

And recent findings in neuroscience suggest that we have more power than we'd ever imagined both to gain courage from others and to give it. When we are only observing another's actions, it turns out that "mirror neurons" in our brains fire as if we were actually performing the observed actions ourselves.[46]

The implications of this finding are stunning. From the simple act of our walking into a grocery store with cloth bag in hand to the gutsy dignity of Egyptians peacefully protesting in Tahrir Square, our actions are actually experienced by, and therefore shaping, anyone observing us.

> We are not individuals who form relationships. We are social animals . . . who emerge out of relationships.
>
> —David Brooks,
> *New York Times* columnist,
> March 7, 2011

So we can consciously use our knowledge of mirror neurons to gain courage. Since we form ourselves in large part through the people we choose to bring into our lives, we can choose our buddies well. We can choose to associate with those more courageous than we. By observing our everyday heroes, we experience courageousness and perhaps become gutsier ourselves. Pretty amazing.

But when Tyler, the young plastic-bottle resister at DePauw, heard me make this point, he smiled. "Don't forget, it's not just other people who are watching. We're watching ourselves too," he told me.

So true. When we observe ourselves taking a stand for what we believe in, doing something risky or just new, our view of ourselves changes forever. Throughout my life, every time I've made a change that aligns with the world I want—whether it's nervously going door to door in a political campaign or choosing a planet-friendly diet—I feel a bit more convincing to myself. A bit more powerful. A bit more of a believer that we all can change. For, think about it. How could we ever believe "the world" can change if we don't experience ourselves changing? And that involves risk—and therefore fear.

An executive coach on trust and fear recently explained his technique for putting fear in its place. I like it. Robert Porter Lynch of the Warren Company, sitting next to me and talking fast on a short flight from Canada, told me that he asks clients to draw two circles, one the size of their fear, the second the size of their passion. "Then I tell them, 'As long as your passion circle is bigger than your fear circle, you're fine. You can achieve what you want.'"[47]

I believe we can trust that as we fuel our passion—as we go with it and grow it—our fears will recede in direct proportion.

## RETHINKING POWER

As you now know, over time I've become more and more convinced that the frames we hold are potent. They determine, quite literally, what we see, what we cannot see, and therefore what we believe is possible. If I'm right that building our civil courage and creative power is critical for our species in this moment, then, as with fear, it's probably smart to ponder our ideas about power too.

The fate of Easter Islanders is a cautionary tale. They became an icon of environmental stupidity after Jared Diamond recounted in *Collapse* how wiping out their own forests brought down their civilization.[48] Historians ferociously debate the many forces, including invasion and disease, at work in their demise, but Easter Islanders were probably not ecologically dumb in any simple sense. Their reigning ideas did, though, give them other priorities than saving trees: One was their idea of power, which islanders long defined and upheld through the creation of, over time, almost nine hundred giant—some of eighty tons or more!—stone statues. Building racks to move these humongous symbols of strength likely required a lot of wood.

So Easter Islanders' idea of power must have been at least one key element doing them in. And for a long time, I've been worried about our ideas about power too, worried that the prevailing environmental frame distracts us from focusing on relationships of power and feeds our ultimate disempowerment—despair, the one luxury we surely cannot afford.

I'm worried that too often our very idea of power is disempowering.

Most of us have absorbed the notion that power is bad. Hear "power," and we immediately think *coercion, force, elitism, domination*. But it all depends. "Power," in its Latin root, means simply the capacity to act. If, to turn our planet toward life, we need more and more of us to exert power even more energetically and thoughtfully, but we are turned off by the very concept of power . . . then that's a problem.

So let's rethink.

What is power? Power is creating. It is making things happen. And it is perhaps the most underappreciated human need. Most experts on what we need to be happy, beyond physical needs, stress the importance of satisfying personal relationships, security, and meaning. But Erich Fromm gets it right in my view. Of humanity he writes,

> [Man] is driven to make his imprint on the world, to transform and to change, and not only *to be* transformed and changed. This human need is expressed in the early cave drawings, in all the arts, in work, and in sexuality. All these activities are the result of man's capacity to direct his will toward a goal and to sustain his effort until the goal is reached.

And if we can't make our "imprint" positively, says Fromm, we try something else: "If . . . man is not able to *act* [he attempts] to restore his capacity. . . . One way is to submit to and identify with a person or group having power. . . . The other way . . . [is] to destroy."[49]

So the question for our species becomes, *Can we consciously reframe this deep need for agency in ways that align with the laws of life? Can we shift from control as the primary expression of power and experience power as co-creating with nature?*

Co-creating power takes discipline. And I don't simply mean remembering to do our part by turning out the lights. I mean mental discipline— continually examining our own assumptions and refusing to go on repeating messages that make us feel virtuous, or release our anger, but ultimately don't work. It means staying open and keeping in touch with our inner two-year-old's "why?"

Such creative power is expanding in part through largely unseen but growing citizen organizations, including Kentuckians for the Commonwealth (KFTC), whose 5,000 plus members take on challenges from toxic dumping to making government more open; in 2010, they were key to halting a new coal plant in their state. Jean True, a KFTC leader the 1990s, told me, "I was home raising kids for ten years. I didn't know anything about politics. I thought my only job was to vote."

When I asked Jean to tell me why she joined KFTC, she responded, "It's just the fun! That you can get together some regular people, go to the capitol, and make changes in state policy. . . . We have a great time going toe to toe and head to head with state legislators. We sometimes know more than they do! It's the fun of power—the ant knocking over the buffalo."

On the other side of the world in the year 2000, I danced with women in a Kenyan village, feeling the exuberant happiness in their newfound power as village tree planters—part of the Greenbelt Movement that has planted 40 million trees.

So, accomplishing humanity's biggest-ever feat may just depend on our shedding the myths now stripping us of power and, with the exuberance of dancing Kenyan tree planters, joining in Jean True's kind of power.

We can do it.

We can do it if we do the work, the good work of checking our failing frames at the door and walking into a new space. There, in our ecological home, everything is connected, so there's no possibility of throwing something "away"—whether a greenhouse gas or the plastic bag CVS just tried to hand me.

Equally true, within an ecological worldview, we're all implicated. The notion of power as a divisible pie—"if he has it, I don't"—dissolves and we see that no one is without power. Even entities as vast as the US Pentagon or as wealthy as Exxon, with its $1,400-a-minute profit, are being shaped moment to moment through our own assumptions about what's possible and by our daily roles in this magnificent drama.

So we have no choice about whether to change the world. We are changing it every day. The choice is only whether our acts contribute to the world we want . . . or not.

Recognizing our power, we can be tough. We can challenge any message telling us we can't have real democracy, so just give up on government, or scolding us for being greedy or apathetic, when most of us want to be part of the solution. We can imagine ourselves like those in the Leipzig church who defied the Stasi, and even like their fellow Germans in Berlin, who—spurred on by their countrymen's courage—started chipping away at the Wall itself.

Sure, I realize that it may be even tougher to perceive and bring down the thought walls that blind us to solutions and freeze us in despair. Yet, as we take out our chisels, we may experience a surprise similar to what Markus's family felt that fateful night in Leipzig—that as we ourselves step out, our allies just keep multiplying.

And with them, in this everyday rethinking of assumptions that no longer fit reality, perhaps we will discover that it is finally, in this century's very challenge to life itself, that humanity comes to realize its own true nature. And comes to see that we are indeed part of nature, not alone, and even to accept and trust ourselves—knowing that we've evolved precisely the capacities we need now, along with ever-greater clarity on the conditions essential to set them free.

What a huge relief.

Together we can create ways of framing our struggle that align with nature, including our own, so as to speed an evolving and immensely liberating ecology of hope. And then, who knows, we may soon come to perceive anything less as a betrayal of *the* distinguishing trait of our species—our power to imagine and therefore to create.

> It will take big, creative imaginations for us to evolve to the next step. Imagination is not fed by fear, but by beauty. . . . Greater than our knowledge of a thing, is a sense of wonder. And, out of that wonder, curiosity, and from that curiosity, a seed of creation.
>
> —Susan Osborn, singer and songwriter[50]

# Cultivating the EcoMind: Moving into Action

## THOUGHT LEAPS FOR THE ECO-MIND

*Thought Trap 1: Endless growth is destroying our beautiful planet, so we must shift to no-growth economies.*

**Thought Leap 1:** Since what we've been calling "growth" leads to so much waste, let's call it what it is—an economics of waste and destruction. Then let's probe *why*. Dropping the distracting "growth-versus-no-growth" debate, we can embrace qualitative notions of where we want to go, choosing terms like "flourishing," and "genuine progress" that focus our minds on enhancing health, happiness, ecological vitality, resiliency, and the dispersion of social power.

*Thought Trap 2: Out-of-control shopping is overtaxing natural resources.*

**Thought Leap 2:** Let's not confuse symptom with cause. "Consumerism"— too much stuff!—is largely a symptom of forces denying us choice. Let's go for *real* choice, as we delink what's become an automatic association between buying things, on the one hand, and enjoying luxury, comfort, and satisfaction, on the other. Together let's imagine and create real luxury— rich, stimulating, and beautiful lives honoring the laws of nature.

*Thought Trap 3: We've had it too good! We must "power down" and learn to live within the limits of a finite planet.*

**Thought Leap 3:** Because most people know they weren't invited to the Too Good party, the message of limits falls flat. An effective and ecologically attuned goal is not about more or less. Moving from fixation on quantities, our focus shifts to what brings health, ease, joy, creativity—more life. These qualities arise as we align with the rules of nature so that our real needs are met as the planet flourishes.

*Thought Trap 4: Humans are greedy, selfish, competitive materialists. We have to overcome these aspects of ourselves if we hope to survive.*

**Thought Leap 4:** Sure, we can be awful. But here's the key to our future: We have also evolved deep capacities for cooperation, empathy, fairness, efficacy, meaning, and creativity. We can't change human nature, but that's OK. We *can* change the norms and rules of our societies to keep negative human potential in check and to elicit these powerful, positive qualities we most need now. Let's focus there with laser intensity.

*Thought Trap 5: Because humans—especially Americans—naturally hate rules and love freedom, we have to find the best ways to coerce people to do the right thing to save our planet.*

**Thought Leap 5:** The nature of nature is rules, and humans love rules that enhance our sense of belonging and that give structure and meaning to our lives. They work and take hold quickly when people feel engaged in shaping them and when they make sense to us. Knowing all this, we can go beyond rules that limit harm and establish ground rules that avoid harm to begin with.

*Thought Trap 6: Now thoroughly urbanized and technology-addicted, we've become so disconnected from nature that it's pretty hopeless to think most people could ever become real environmentalists.*

**Thought Leap 6:** We can count on one thing for sure: Humans evolved to love nature. So city dwellers, too, can and are rediscovering this natural connection, finding that it enhances our health, learning, and fun. As we reconnect with nature, it's likely that support for public action to heal the natural world will grow too.

*Thought Trap 7: It's too late! Human beings have so far overshot what nature can handle that we're beyond the point of no return. Democracy has failed—it's taking way too long to face the crisis. And because big corporations hold so much power, real democracy, answering to us and able to take decisive action, is a pipe dream.*

**Thought Leap 7:** It's not too late for life. And its flourishing depends on democracy attuned to all we now know of our nature and with what history teaches: that centralized power cannot lead to the sustained good of the whole. Plus, today's crises are so interrelated, vast in scope, and complex that solutions depend on tapping the direct experience, common sense, and passion of people most affected. And that's billions of us! With the buy-in that comes with having a real voice, most of us want to contribute to solutions that turn our planet toward life. So the "mother of all issues" is removing the power of concentrated wealth over public decision making and infusing voices of empowered citizens in all arenas of public life. It's beginning in surprising places.

## LEXICON FOR THE ECO-MIND

> *If we wish to renew our solidarity with the sensuous earth,*
> *then we shall have to learn to speak in some new ways . . .*
> *to speak more in accordance with our animal senses.*

—David Abram, "The Invisibles," *Parabola Magazine,* 2006

- From staying within the limits of nature to aligning with nature
- From economies of waste and destruction to economies enhancing life
- From freedom as "get out of my way" to freedom as power in a Living Democracy
- From privately held government to publicly accountable government
- From one-rule economies, leading inexorably to unfair concentration, to economies kept open and fair via democratically chosen rules
- From "free market" as the free rein of corporations monopolizing power to "free market" as the freedom of all to participate in the market
- From the assumption of scarcity of goods and goodness to the premise of possibility
- From debating the goodness *of* human nature to recognizing the goodness *in* human nature to be tapped for positive ends
- From overcoming human nature to aligning societies' rules with human nature—the good, the bad, and the ugly
- From separateness in which we're either givers or receivers to connectedness in which we're always both

- From the "blame game" to the assumption of mutual accountability
- From nature as merely divisible property to nature as a commons in our common care
- From power as control to power as our capacity to co-create
- From "random acts of sanity" to conscious acts building our democratic power
- From democracy as only a structure of government—Thin Democracy—to democracy as a rewarding way of life—Living Democracy
- From citizenship as simply the act of voting to citizenship as the satisfying lifelong practice of the arts of democracy
- From calling on others to become "better" people to encouraging and enabling oneself and others to be more creative and courageous
- From paralyzing despair to honest hope

## UPPING OUR CIVIL COURAGE: THE POWER OF PUTTING OUR ECO-MINDS INTO ACTION

### Some Questions to Explore with Others

Don't read this book alone!

*The power of connection*: Who can I reach out to right now—friends, strangers, groups—to help keep me going? (In the following pages you'll find brief introductions to organizations doing great work as well as sources for ongoing learning and inspiration.)

*The power of curiosity*: What have I just learned that most piques my curiosity and inspires me to learn more? How can I best pursue my curiosity?

*The power of frame and language*: What is one piece of my current mental map—my core assumptions about life—that limits me? How could I reframe it to free myself? What words do I use that keep my thinking mired in the world of separateness and lack? What are other terms I want to start using?

*The power of action*: What is one thing I learned in *EcoMind* that I want to act on right now to align my life with the world I want and make me more powerful?

*The power of agitation*: What most disturbed me in this book? How can I transform that disturbance into energy for action?

*The power of self-awareness*: What is an important strength I already have—knowledge, contacts, quality of my character—that I can share and further the emergence of Living Democracy? How do I grow my strengths? How do I use my strengths to empower others?

*The power of daily attention*: What is one change I can adopt right now that enriches my life because it reminds me every day of my embeddedness in nature?

*The power of inspiration*: Who are my heroes—my everyday heroes—and how can I bring them more fully into my life?

*The power of embracing fear*: What is one risk I could take now to enhance my creative power?

*The power of organization*: How can I incorporate eco-mind lessons into groups I'm part of? How can I join and strengthen other groups (perhaps among those described below) that are aligned with a frame of possibility?

Please share your thoughts and get more ideas at www.smallplanet .org.

## MAKING OUR OWN CONNECTIONS

Earlier I stressed the importance of "buddies," co-learners as we develop our eco-minds. So the Small Planet Institute has developed a simple, fun workshop that could be adapted to an informal gathering of friends, a classroom, or a conference. To create a new group, you might consider using Meetup Everywhere, part of Meetup.com (for which, full disclosure, my partner Richard Rowe consults). It's an online tool we've used to spark face-to-face gatherings. Please visit smallplanet.org to access our workshop guide (including video clips) for translating eco-mind ideas into action, share feedback on your group's conversations, and find links to other great organizations.

## Connecting with organizations and campaigns that are . . .

Here is a start. Enjoy!

### Engaging Citizens in the Work of Planet Healing

**350. org.** Founded by renowned author Bill McKibben, 350.org works to build a global movement of the scale needed to meet the climate challenge. The number 350 refers to the level of parts per million of $CO_2$ in our atmosphere that many scientists identify as the safe limit. In 2010, 350.org inspired more than 7,000 climate-solution events in 187 countries. www.350.org

**Alliance for Climate Education.** Since 2009, the Alliance for Climate Education has reached over 1.3 million students at over 1,200 schools across the country, inspiring them to form hundreds of clubs to take on climate projects. The University of Chicago Lab School Action Team, for example, organized its community to recycle over 22,000 pounds of e-waste. My godson Matt Lappé helps to lead this fantastic outfit. www.acespace.org

**American Sustainable Business Council.** Partners in this advocacy coalition are over 150,000 businesses and 300,000 business executives, owners, investors, and others—all working for public policies that support a vibrant, just, and sustainable economy. It advocates, for example, expansion of green jobs and an end to corporate tax-haven abuse. www.asbcouncil.org

**Business Alliance for Local Living Economies.** Featured in Thought 1 as a leader in democratizing economic life, BALLE has grown to 30,000 innovators, 80 local business networks, 450,000 local jobs in the United States and Canada. BALLE supports "Local First" campaigns to vitalize communities by strengthening socially responsible, independent businesses. http://bealocalist.org

**Campus Climate Challenge.** Connecting more than thirty leading youth organizations across North America, Campus Climate Challenge works for 100 percent clean-energy policies in high schools and on college campuses. During its 2007 Campus Climate Challenge, a week of events on nearly six hundred campuses in forty-nine states and eight Canadian provinces, more than 50,000 students spoke out for action on climate change. www.climatechallenge.org

**Center for Ecoliteracy.** The Center for Ecoliteracy helps educators apply ecological thinking and practice from English classroom to garden. It advances education for sustainable living through books and teaching guides—many available on its website—as well as professional development seminars, a sustainability leadership academy, keynote presentations, and consulting. Its pioneering work has promoted school gardens and school lunches. www.ecoliteracy.org

**Cloud Institute for Sustainable Education.** Works with K–12 schools to develop and spread learning materials and teaching approaches that inspire and prepare young people to shape a society committed to sustainable development. www.cloudinstitute.org

**Corporate Accountability International.** For nearly thirty-five years, Corporate Accountability International (formerly Infact) has run highly effective campaigns to save lives, protect public health, and preserve the environment. Its campaigns have compelled dramatic changes in corporate conduct, from curbing the life-threatening marketing of infant formula in the developing world to securing strong new global protections against the marketing of tobacco products to children. The organization works to protect water resources, guarantee universal access to clean drinking water, and promote the long-term viability of public water systems. Its Challenging Corporate Abuse of Our Food campaign is helping to reverse the deadly epidemic of diet-related disease. www.stopcorporateabuse.org

**Earth Circles.** Earth Circles are six to ten engaged citizens who come together to share their environmental concerns in small, self-facilitated groups in their neighborhoods, schools, communities, workplaces, and faith groups. The Earth Circles workbook offers a series of concrete suggestions for participants to focus on the environmental challenges of their locale and translate their sustainability concerns into empowering action. www.earth-circles.org

**EcoLogic Development Fund.** Featured in the final chapter for its agroecological work in Guatemala, this empowering organization works with communities to foster sustainable livelihoods while at the same time protecting biodiversity. In Central America and Mexico, it helps poor, rural communities take charge of managing local natural resources, enabling local partners to become environmental stewards of the land. www.ecologic.org

**Food Democracy Now!** Food Democracy Now! is a grassroots organization, launched in 2008, to address our failing food system. Its aim: equal access to healthy food and respect for the dignity of the farmers and farm workers producing food sustainably. In December 2010, Food Democracy Now! delivered more than 200,000 public comments to the Department of Justice and USDA asking that they end monopoly abuses in food and agriculture. www.fooddemocracynow.org

**Friends of the Earth.** Since 1969, these progressive environmental activists have promoted clean energy and solutions to climate change, as well as safe and sustainable food and healthy marine ecosystems. Friends of the Earth has prevented 150 destructive dam and water projects; its campaigns have resulted in an increase of $17 billion for public transportation, and $50 billion less in government spending for nuclear facilities. www.foe.org

**Grassroots International.** Understanding "development" as partnership, for three decades Grassroots has worked shoulder to shoulder with initiatives of peasants, family farmers, women, and indigenous groups through grant making, education, and advocacy. It works internationally and with US activists, allies, and donors to build a global movement for the human right to land, water, and food and to remove obstacles in the path of people working for food democracy and dignity. www.grassrootsonline.org

**Green for All.** A national organization, Green for All works to build the green economy, with equal access to green jobs for all. In 2009, a Green for All–assembled coalition ensured that clean-energy opportunities for vulnerable communities were in the Climate Bill passed by the U.S. House of Representatives. www.greenforall.org

**Local Harvest.** This great website connects individuals with local farmers' markets, community gardens, family farms, and other sources of sustainably grown food. Type in your zip code to find the wealth of local, healthy food resources near you. www.localharvest.org

**Natural Step.** Since 1989, the Natural Step has been a pioneer in helping businesses and institutions—from Nike and IKEA to the US Marines—and communities integrate sustainability principles into their core strategies, decisions, and operations. It provides education and training to build organizational capacity for sustainable solutions. www.thenaturalstep.org and www.thenaturalstep.org/en/usa

**PowerShift/Energy Action Coalition.** Yet another movement fueled by the power of youth, this coalition of more than fifty youth-led environmental and social justice organizations works for "bold action on climate and energy" *now*. It has convened three national youth climate summits, known as PowerShift, bringing as many as 10,000 young leaders to Washington, DC, for common action. www.wearepowershift09.org

**Rainforest Action Network (RAN).** A national leader addressing many of this book's themes—from global warming to rain forest protection, from clean energy and human rights to sustainable economies. It got twenty fashion giants, for example, Gucci Group, Tiffany & Co., Hugo Boss, and H&M, to agree to stop buying paper from Asia Pulp and Paper, a company responsible for a colossal share of rainforest loss in Indonesia. Anna Lappé is a member of the board. www.ran.org

**Real Food Challenge.** This exciting effort is working to leverage the purchasing power of university campuses to help generate a just and sustainable food system. Its goal is to increase the amount of "green" foods purchased by colleges to 20 percent by 2020. Launched in 2009, it has already brought over three hundred schools on board. It succeeded in replacing Panda Express with a student-run co-op at the University of California, Berkeley. www.realfoodchallenge.org

**Jane Goodall's Roots & Shoots.** Initiated in Tanzania with twelve high school students twenty years ago, this movement has grown into a global environmental and humanitarian network involving tens of thousands of young people of all ages in more than 125 countries. It is youth-driven and members are empowered to choose projects and take action to make the world a better place for people, animals, and the environment. They acquire leadership skills, develop a global perspective, and seek ways to promote peace with each other and the natural world. www.rootsandshoots.org

**Sierra Club.** With over 1.4 million members, the Sierra Club is the oldest and largest US environmental organization. It works to reverse global warming and increase our renewable energy. In 2011 in Illinois, it helped achieve the veto of two bills that would have forced natural gas consumers to bear the expenses from two new coal plants that process coal into a synthetic form of natural gas. www.sierraclub.org

**Slow Food.** Slow Food boasts 100,000 members in 150 countries "helping people understand the importance of caring about where their food comes from, who makes it and how it's made." In the US, Slow Food has partnered with schools to bring fresh and organic foods to students from kindergarten to college. The Small Planet Institute teamed up with Slow Food to create fact-filled "table tents" college students and others can use to share key facts about food-related energy and climate challenges. www.slowfood.com

**Transition Movement.** Across the globe, roughly four hundred "official" Transition Initiatives, and many more affiliated communities, are part of the Transition Movement, a vibrant, grassroots effort that began in the UK in 2006. Taking root from New Zealand and South Africa to the US, it is uniting citizens to transform the climate and energy challenges into community life that is more abundant, fulfilling, equitable, and socially connected. Does your town have a Transition Initiative? www.transitionnetwork.org and www.transitionus.org

**World Social Forum and US Social Forum.** The World Social Forum is an open meeting drawing hundreds of thousands of people together annually since 2001. Social movements, networks, and civil society organizations share their thinking and ideas about solutions to the world's ecological and economic crises. Two US Social Forums—in 2007 and 2010—gathered tens of thousands. May they continue. www .forumsocialmundial.org and www.ussf2010.org

> *Find new action pals*: As noted above, through online tools like Meetup.com, you can announce your interest and meet others for learning and action.

## Working to Involve Citizens and Remove the Power of Private Wealth in Our Democracy

**Americans for Campaign Reform.** A bipartisan initiative and a nonprofit group, Americans for Campaign Reform works to achieve voluntary public funding of national elections. It's helped further citizens' and legislators' support for the Fair Elections Now Act, which gained many co-sponsors in the House in 2010. www.acrreform.org

**Coffee Party.** Founded in 2010, the Coffee Party is creating, through its chapters, welcoming meeting places for those with diverse backgrounds and views to join in common action to strengthen our democracy. Participants are encouraged to make a "civility pledge" so that communication remains respectful. The Coffee Party actively involves its participants in deciding on the focus of its work, and overwhelmingly Coffee Partiers have chosen as a top priority getting money out of politics via public financing. www.coffeepartyusa.com

**Common Cause.** For decades Common Cause has served as a watchdog to ensure that government supports the common good. It helped to bring about the creation of the independent Office of Congressional Ethics, which leads investigations in response to complaints or concerns about House member conduct. www.commoncause.org

**DEMOS.** Through research, publishing, and advocacy, DEMOS works to influence public debate toward creating a more inclusive and vibrant political democracy and economy. www.demos.org

**Public Campaign.** You can turn to this nonpartisan group to find out what is happening on implementing voluntary public financing ("clean" elections) in your state. Since the late 1990s, Public Campaign has helped more than a dozen states, including Arizona, Maine, New Jersey, North Carolina, and Oregon, achieve cleaner and fairer elections. www.publicampaign.org

**Represent.Us.** This bi-partisan campaign works to make elections and campaign finance transparent, accountable, and fair. To this end, it is championing the American Anti-Corruption Act, which if passed would impose strict lobbying and conflict of interest laws, end secret political money, and enable candidates to run without big-money backers. http://represent.us and http://anticorruptionact.org

## Engaging Citizens in Reclaiming Our Democracy
### Groups Creating New Ways to Empower Citizen Voices

**America***Speaks.* This nonpartisan, nonprofit organization offers proven methods for large-group discussion and decision making in which all can feel heard without manipulation or bias. America*Speaks* has helped more than 160,000 citizens make tough choices and become involved in problem solving in their communities. www.americaspeaks.org

**Center for Deliberative Democracy.** The Center for Deliberative Democracy, housed in the Department of Communication at Stanford University, developed and promotes Deliberative Polling, a process in which a representative sample of citizens discusses a critical issue, with the help of trained moderators and carefully balanced briefing materials. Before and afterward, a poll is conducted to show how the process affected the participants' views on public issues. http://cdd .stanford.edu

**Co-Intelligence Institute.** Co-intelligence is founded on wholeness, collaboration, and creativity. The institute applies these values to democratic renewal and meeting national and global crises to achieve sustainable cultures. www.co-intelligence.org

**National Coalition for Dialogue and Deliberation.** With fourteen hundred organizations and individuals, NCDD promotes collaborating among practitoners, leaders, scholars, and groups, using dialogue and deliberation to tackle society's challenges. ncdd.org

**Everyday Democracy.** This leader in community problem solving enables people of all backgrounds and views to talk together in facilitated study circles so they can work through issues and take action. Since 1989, Everyday Democracy has worked with more than six hundred communities on challenges ranging from racial inequality to poverty to violence. www.everyday-democracy.org

## Groups Focusing on Cultivating the Arts of Democracy

**Center for Nonviolent Communication.** An international training and consulting group offering an approach to communication that enables people to "peacefully and effectively resolve conflicts in personal, organizational, and political settings." Active in over sixty-five countries, it seeks to develop more sustainable, compassionate human interactions, both personally and professionally. www.cnvc.org

**Community at Work.** A consulting group that "puts participatory values into practice," Community at Work facilitates group decision making and studies the dynamics involved in reaching sustainable agreements.

The organization's *Facilitator's Guide to Participatory Decision-Making* and its training helped those creating the student-led sustainability course at University of California, Santa Cruz, mentioned in Thought 4. www .communityatwork.com

**Dynamic Governance (DG)/Governance Alive.** Here you can learn about a proven, practical method of inclusive, democratic decision making in complex organizations. Created in the Netherlands in the 1970s, DG is now proving effective not only in businesses but also in human-development organizations, cohousing communities, and families. www .governancealive.com

## SOME FAVORITE READING: BOOKS, MAGAZINES, AND WEBSITES FOR THE ECO-MIND

### Books

Benyus, Janine M. *Biomimicry: Innovation Inspired by Nature*. New York: Harper Perennial, 1997.

Duval, Jared. *Next Generation Democracy*. New York: Bloomsbury, 2010.

Fromm, Erich. *The Anatomy of Human Destructiveness*. New York: Holt, Rinehart & Winston, 1973.

Girardet, Herbert, and Miguel Mendonça. *A Renewable World: A Report for the Future World Council*. Totnes, UK: Green Books, 2009.

Hawken, Paul, Amory B. Lovins, and L. Hunter Lovins. *Natural Capitalism*. Rev. ed. Boston: Little, Brown and Company, 2010.

Hester, Randolph. *Design for Ecological Democracy*. Cambridge, MA: MIT Press, 2006.

Hrdy, Sarah Blaffer. *Mothers and Others: The Evolutionary Origins of Mutual Understanding*. Cambridge, MA: Belknap Press, 2009.

Jones, Van. *The Green Collar Economy: How One Solution Can Fix Our Two Biggest Problems*. New York: HarperCollins, 2008.

Lappé, Anna. *Diet for a Hot Planet: The Climate Crisis at the End of Your Fork and What You Can Do About It*. New York: Bloomsbury, 2010.

Solnit, Rebecca. *Hope in the Dark: Untold Stories, Wild Possibilities*. New York: Nation Books, 2004.

Stone, Michael. *Smart by Nature*. Healdsburg, CA: Watershed Media, 2009.

Von Weizsäcker, Ernst, Amory B. Lovins, and L. Hunter Lovins. *Factor Four: Doubling Wealth, Halving Resource Use*. London: Earthscan, 1998.

Walljasper, Jay. *All That We Share*. New York: New Press, 2010.

## Magazines

*Adbusters* (Canada): www.adbusters.org

*New Internationalist* (UK): www.newint.org

*Ode* (Netherlands): www.odemagazine.com

*Resurgence* (UK): www.resurgence.org

*Solutions* (US): www.thesolutionsjournal.com

*Yes!* (US): www.yesmagazine.org

## Websites

BlueEconomy.de (in Germany)

ClimateChangeNews.org

Commondreams.org

EWG.org (Environmental Working Group)

FutureWeWant.org

Green.tv (in the UK)

Grist.org

MakeMeSustainable.com

OntheCommons.org

PlanetChange.tv

TreeHugger.com

WorldFutureCouncil.org

YourOliveBranch.org

# Acknowledgments

In today's lingo, *EcoMind* is truly "crowd-sourced," so a crowd must surely be thanked. Below are listed many of the generous readers whom I've never met but who, via our online forum, contributed useful and provocative ideas to the first draft of this book. My gratitude begins with you, for your responses convinced me of the importance of this project.

Now to those I am fortunate to know. First, to my wise and supportive partner, Richard Rowe, for the endless conversations shaping this book that always challenge my thinking; and for your faith in me that is a daily source of strength and happiness; and to my children, Anna and Anthony Lappé, and their spouses, Clarice Lappé and John Marshall, for your loving support on this journey.

Next, I am grateful for the keen insights and tireless work of friends and coworkers at the Small Planet Institute, especially Brooke Ormond and Samantha Mignotte. As institute manager, Brooke expertly organized all manner of support to move this book to completion; and Samantha, with grace and intelligence, oversaw the lengthy research. You were steadfast and indefatigable throughout. What a gift you have been. Arriving at Small Planet in the last weeks of this project, Danya Rumore deserves my thanks for her enthusiasm that helped push us over the finish line.

Others who added enormously in the final stages are Douglas Dupler and Martin O'Hare. With thanks for offering your expertise on specific content to David Thompson, Maja Goepel, Jim Benedix, Tim LaSalle,

Ann Adams, Miguel Mendonca, and Richard Wallace. For the initial draft, two people deserve very special thanks: Andrea Diehl, a gifted writer and friend who contributed expert editorial assistance; and Jessica Bruce, for months of invaluable research. Other excellent researchers were Katherine McDonald, Alex Tung, Nicole Crescimanno, and Durrie Bouscaren.

For your supportive friendship, thank you Linda Pritzker, Jan Surrey, Steve Bergman, Hathaway Barry, Aaron Stern, Susan Kanaan, and Josh Mailman. Friends Carolyn Kniazzeh, Dorothy Stoneman, Hannes Lorenzen, Mishy Lesser, Sarah Conn, Nomi Rowe, Loren Stein, and Bruce Haynes, as well as seven volunteer reviewers—Danielle Connor, Dean Cycon, Dave Forrest, Jessica Lotak, Eric Zamost, and Bruce Wilkinson—offered important feedback on the initial draft, convincing me to move the project forward. Also, thanks to Julie Jensen, who organized our online outreach to make the "crowd-sourcing" possible.

In addition, a big thank-you goes to my savvy agent Deirdre Mullane and my encouraging and perceptive partners at Nation Books, especially my editor Ruth Baldwin, publicist Andrea Bussell, production manager Sandra Beris, and copyeditor Jen Kelland Fagan.

Finally, before I acknowledge many of the online and other contributors of feedback who shaped this book, let me state what I hope is obvious: While I am grateful for all the help I received, responsibility for errors of thought and fact lie exclusively with me. And I am eager to know what they are. So please take advantage of the online *EcoMind* forum at smallplanet.org to offer your suggestions on any aspect of this book—and especially, I hope, on how to take its ideas into action.

| | | |
|---|---|---|
| A. Alan | Pat Evans | Chuck Kleymeyer |
| Elsa Bengel | Julie Fanselow | Ben Kunesh |
| John Bengel | Aquene Freechild | Paul LeVie |
| Isabel Best | John Gershman | Dick Levins |
| Daniel Christopher | Lucia Goyen | Bruce Macomber |
| Daniel Cohen | Amanda Hagood | Tim Magner |
| David K. Cundiff | Rebecca Halleran | Leslie Marshall |
| Jesse de la Rosa | Kate Horner | Nicholas Rooney Martins |
| Ann Delorey | Peter Kelley | Joan McVilly |

Beth Norcross

Alan Page

Crossley Pinkstaff

Wendell Refior

Phil Ritter

Scott A. Root

Joe Rosenfeld

Helen Sitar

Maryhelen (Mel) Snyder

Amy Southworth

Marcy Square

Steve Stodola

Clint Stretch

JJ Tiziou

Jim Tolstrup

Jennifer Troobnick

Greg Vaughan

Frances Moore Lappé
*Cambridge, Massachusetts, June 2011*

# Notes

## Why I Wrote This Book

1. Tom Crompton and Tim Kasser, *Meeting Environmental Challenges: The Role of Human Identity* (London: WWF-UK, 2009), 18. Available at http://assets.wwf.org.uk/downloads/meeting_environmental_challenges___the_role_of_human_identity.pdf.

## Our Challenge

1. "Past Climate Change," Environmental Protection Agency, www.epa.gov/climatechange/science/pastcc.html (last modified September 8, 2009); David Chandler, "Climate Change Odds Much Worse Than Thought: New Analysis Shows Warming Could Be Double Previous Estimates," *MIT News*, May 19, 2009. http://web.mit.edu/newsoffice/2009/roulette-0519.html.

2. Justin Gillis, "As Glaciers Melt, Science Seeks Data on Rising Sea Levels," *New York Times*, November 11, 2010. Available at http://www.nytimes.com/2010/11/14/science/earth/14ice.html. W. Tad Pfeffer, J. T. Harper, and S. O'Neel, "Kinematic Constraints on Glacier Contributions to 21st-Century Sea-Level Rise," *Science* 321, no. 5894 (September 2008): 1340–1343; Martin Vermeer and Stefan Rahmstorf, "Global Sea Level Linked to Global Temperature," *Proceedings of the National Academy of Sciences of the United States* 106, no. 51 (December 2009): 21527–21532. www.pnas.org/content/106/51/21527.full.pdf+html.

3. "Indus River Overflows 40 Times Its Volume of Water," BBC News, August 28, 2010, www.bbc.co.uk/news/world-south-asia-11120483.

4. John Vidal, "Australia Suffers Worst Drought in 1,000 Years," *Guardian*, November 8, 2006, www.guardian.co.uk/world/2006/nov/08/australia.drought; J. David Goodman, "Australia Floods Show No Signs of Retreating," *New York Times*, December 31, 2010, http://www.nytimes.com/2011/01/01/world/asia/01australia.html.

5. Food and Agriculture Organization (FAO) of the United Nations, *Global Forest Resources Assessment 2010* (Rome: FAO, 2010), 15–23. www.fao.org/docrep/013/i1757e/i1757e.pdf; Arild Angelsen et al., *Reducing Emissions from Deforestation and Forest Degradation (REDD): An Options Assessment Report* (Washington, DC: Meridian Institute, 2009), 1. www.redd-oar.org/links/REDD-OAR_en.pdf.

6. James W. Kirchner and Anne Weil, "Delayed Biological Recovery from Extinctions Throughout the Fossil Record," *Nature* 404 (March 2000): 177–180.

7. US Energy Information Administration, *International Energy Outlook 2010* (Washington, DC: Energy Information Administration, 2010), 124, 134. www.eia.doe.gov/oiaf/ieo/pdf/0484%282010%29.pdf.

8. International Labor Organization, *Global Unemployment Trends 2012*, http://www.ilo.org/wcmsp5/groups/public/@dgreports/@dcomm/@publ/documents/publication/wcms_171571.pdf; World Food Situation, Food Price Index, UN Food and Agriculture Organization. Compared with an index of 100, the average for 2002–2004, the Food Price Index in 2011 remained above 200 for most of the year. http://www.fao.org/worldfoodsituation/wfs-home/foodpricesindex/en/

9. Andrew Simms, "Planet Crunch," *Resurgence* 253 (March–April 2009), www.resurgence.org/magazine/article2746-planet-crunch.html.

10. Jeffrey M. Jones, "In U.S., Concerns About Global Warming Stable at Lower Levels," Gallup, March 14, 2011, www.gallup.com/poll/146606/Concerns-Global-Warming-Stable-Lower-Levels.aspx.

11. Erich Fromm, *The Anatomy of Human Destructiveness* (New York: Holt, Rinehart &Winston, 1973), 259–261.

12. Daniel J. Simons and Christopher F. Chabris, "Gorillas in Our Midst: Sustained Inattentional Blindness for Dynamic Events," *Perception* 28 (1999): 1059–1074; Christopher Chabris and Daniel Simons, *The Invisible Gorilla: And Other Ways Our Intuitions Deceive Us* (New York: Crown, 2010).

13. Jules Pretty, *The Earth Only Endures: On Reconnecting with Nature and Our Place in It* (London: Earthscan, 2007), 217.

14. J. R. McNeill, *Something New Under the Sun: An Environmental History of the Twentieth-Century World* (New York: W. W. Norton, 2000), 15.

15. Millennium Ecosystem Assessment, *Ecosystems and Human Well-Being: Synthesis*

(Washington, DC: Island Press, 2005), 4. www.maweb.org/documents/document .356.aspx.pdf.

16. Rachel Cleetus, Steven Clemmer, and David Friedman, *Climate 2030: A National Blueprint for a Clean Energy Economy* (Cambridge, MA: Union of Concerned Scientists, 2009), 74–75. www.ucsusa.org/global_warming/solutions/big_picture_solutions/ climate-2030-blueprint.html.

17. *Common Sense on Climate Change: Practical Solutions to Global Warming* (Cambridge, MA: Union of Concerned Scientists, n.d.), 3. www.ucsusa.org/assets/ documents/global_warming/climatesolns.pdf; data on US oil consumption is from "Oil: Crude and Petroleum Products Explained," US Energy Information Administration, www.eia.doe.gov/energyexplained/index.cfm?page=oil_home#tab2 (accessed January 5, 2011).

18. Amory Lovins et al., *Winning the Oil Endgame* (Snowmass, CO: Rocky Mountain Institute, 2004), xii. www.rmi.org/rmi/Library/E04-07_WinningTheOilEndgame; James Glanz, "The Economic Cost of War," *New York Times*, February 28, 2009, www .nytimes.com/2009/03/01/weekinreview/01glanz.html.

19. Lovins et al., *Winning the Oil Endgame*, ix–xiii; Amory Lovins, "Freeing America from Its Oil Addiction," January 4, 2010, CNN Opinion, http://articles .cnn.com/2010-01-04/opinion/lovins.weaning.us.off.oil_1_plug-in-hybrid-whale-oil -batteries?_s=PM:OPINION.

20. Carole Bass, "Amory Lovins: Energy Efficiency Is the Key," *Yale Environment 360*, November 26, 2008, http://e360.yale.edu/content/feature.msp?id=2091.

21. Renewable Energy Policy Network for the 21st Century (REN21), *Renewables 2010: Global Status Report* (Paris: REN21 Secretariat, 2010), 22. www.ren21.net/ Portals/97/documents/GSR/REN21_GSR_2010_full_revised%20Sept2010 .pdf.

22. "Ecological Debt and Oil Moratorium in Costa Rica," Oil Watch, August 2005, www.oilwatch.org/doc/campana/deuda_ecologica/deuda_costarica_ing.pdf, 11; REN21, *Renewables 2010*, 59; Stefan Lovgren, "Costa Rica Aims to Be 1st Carbon-Neutral Country," National Geographic Daily News, March 7, 2008, http://news .nationalgeographic.com/news/2008/03/080307-costa-rica.html.

23. "$CO_2$ Emissions (Metric Tons per Capita)," World Development Indicators Data, World Bank, http://data.worldbank.org/indicator/EN.ATM.CO2E.PC/countries (accessed February 28, 2011).

24. Janet L. Sawin and William R. Moomaw, "Renewing the Future and Protecting the Climate," *Worldwatch Magazine* 23, no. 4 (July–August 2010): 30.

25. Janet L. Sawin and William R. Moomaw, "Renewing the Future and Protecting the Climate," 30.

26. Angelika Pullen, Liming Qiao, and Steve Sawyer, eds., *Global Wind 2009 Report* (Brussels: Global Wind Energy Council, 2010), 10. www.gwec.net/fileadmin/documents/ Publications/Global_Wind_2007_report/GWEC_Global_Wind_2009_Report_LOWRES _15th.%20Apr..pdf.

27. FAO, *Global Forest Resources Assessment 2010*, xiii, 98, 100, 233, 261–265. www .fao.org/docrep/013/i1757e/i1757e.pdf.

28. Bina Agarwal, *Gender and Green Governance: The Political Economy of Women's Presence Within and Beyond Community Forestry* (London: Oxford University Press, 2010), xx, Chapter 4, 119–120.

29. "Brazil Protects Climate with Record Low Amazon Deforestation," Environmental News Service, December 1, 2010, www.ens-newswire.com/ens/dec2010/2010-12 -01-01.html.

30. Robert J. Diaz and Rutger Rosenberg, "Spreading Dead Zones and Consequences for Marine Ecosystems," *Science* 321, no. 5891 (August 2008): 926–929.

31. "2010 Forecast of the Summer Hypoxic Zone Size, Northern Gulf of Mexico," Louisiana Universities Marine Consortium and Louisiana State University, Hypoxia Forecast, June 28, 2010, www.lumcon.edu/UserFiles/File/LUMCON%20LSU%202010 %20hypoxia%20forecast%20NNR.pdf.

32. Anna Lappé, *Diet for a Hot Planet: The Climate Crisis at the End of Your Fork and What You Can Do About It* (New York: Bloomsbury, 2010), 11, 36–37, for more information, visit the Food Climate Research Network at www.fcrn.org.uk.

33. JINKUN, "African Agricultural Policies and the Development of Family Farms," Genetic Resources Action International (GRAIN), April 23, 2009, www.grain .org/front_files/copagen-april-2009-en.pdf; "Production: Production Indices," FAO-STAT Data Tables, http://faostat.fao.org (accessed April 2, 2011).

34. Catherine Badgley et al., "Organic Agriculture and the Global Food Supply," *Renewable Agriculture and Food Systems* 22 (2007): 86–108.

35. *Foresight Project on Global Food and Farming Futures: Synthesis Report C9: Sustainable Intensification in African Agriculture Analysis of Cases and Common Lessons*, UK Government Office for Science, January 2011, 8. //www.bis.gov.uk/assets/bispartners/ foresight/docs/food-and-farming/synthesis/11-629-c9-sustainable-intensification -in-african-agriculture.pdf.

36. Natural Resource and Environment Department, *Organic Agriculture, Environ-*

*ment and Food Security* (Rome: Food and Agriculture Organization of the United Nations, 2002), www.fao.org/DOCREP/005/Y4137E/y4137e02b.htm#96.

37. Helga Willer, "Current Status of Organic Farming Worldwide," *Ecology and Farming* nr 2 (April 2011): 11, http://ecologyandfarming.com/artikelen/Status-0211.pdf; Kate Trumper et al., *The Natural Fix? The Role of Ecosystems in Climate Mitigation* (Cambridge, UK: UN Environment Program, 2009), 41, 55, www.unep.org/pdf/BioseqRRA_scr.pdf.

38. Robert B. Reich, *Supercapitalism: The Transformation of Business, Democracy, and Everyday Life* (New York: Knopf, 2007), 106.

39. Doug Henwood, ed., Left Business Observer, personal communication with author, February 24, 2011. Calculations based on Bureau of Labor Statistics data on the nominal value of the minimum wage, deflated by the consumer price index, rebased to February 2011 dollars; Mark Greenberg, Indivar Dutta-Gupta, and Elisa Minoff, eds., *From Poverty to Prosperity: A National Strategy to Cut Poverty in Half* (Washington, DC: Center for American Progress, April 2007), 9. www.americanprogress.org/issues/2007/04/pdf/poverty_report.pdf.

40. Charles Blow, "Suffer the Little Children," *New York Times,* December 24, 2010. http://www.nytimes.com/2010/12/25/opinion/25blow.html. For US poverty, see *Current Population Survey, 2010: Annual Social and Economic Supplements* (Washington, DC: US Census Bureau, 2010), Historical Poverty Tables, Table 9. www.census.gov/hhes/www/poverty/data/historical/people.html. For UK poverty, see Nick Adams et al., eds., *Households Below Average Income: An Analysis of the Income Distribution 1994/95–2008/09* (London: Department for Work and Pensions, 2010), 104. http://statistics.dwp.gov.uk/asd/hbai/hbai_2009/pdf_files/fullZ_hbai10.pdf.

41. *Vietnam: Poverty Reduction Strategy Paper–Annual Progress Report* (Washington, DC: International Monetary Fund, 2006), 1–2, http://www.imf.org/external/pubs/ft/scr/2006/cr0670.pdf.

42. Ricardo Barros et al., "Markets, the State, and the Dynamics of Inequality: Brazil's Case Study," Research for Public Policy: Inclusive Development-RBLAC-UNDP, 2009, http://economiccluster-lac.org/images/pdf/desarrollos-Incluyentes/14_RPPLAC_ID.pdf, 24; Kathy Lindert et al., *The Nuts and Bolts of Brazil's Bolsa Familia Program: Implementing Conditional Cash Transfers in a Decentralized Context* (Washington, DC: World Bank, 2007) http://siteresources.worldbank.org/INTLACREGTOPLABSOCPRO/Resources/BRBolsaFamiliaDiscussionPaper.pdf.

43. FAOSTAT Data Tables (Supply Utilization Accounts and Food Balances: Food Balance Sheets), http://faostat.fao.org (accessed March 1, 2011); ETC Group, "Who Will Feed

Us? Questions and Answers for Food and Climate Crises," *Communiqué* 102 (November 2009), 1, www.etcgroup.org/upload/publication/pdf_file/ETC_Who_Will_Feed_Us.pdf; *Towards a Green Economy: Pathways to Sustainable Development and Poverty Eradication* (United Nations Environmental Programme, 2011), 41–42, http://unep.org/Green Economy/Portals/93/documents/Full_GER_screen.pdf.

44. "Noteworthy: Urban Waves of Grain," *Solutions* 1, no. 5 (September–October 2010): 9–10.

45. Gustavo Capdevila, "Dim Outlook Despite Economic Growth, Says ILO," Interpress Third World News Agency, August 23, 1999; Regional Office for Latin America and the Caribbean, *2006 Labour Overview* (Geneva: International Labor Organization, 2006), 13, 32. www.oit.org.pe/WDMS/bib/publ/panorama/labour_o.

46. National Commission for Enterprises in the Unorganized Sector (NCEUS), *The Challenge of Employment in India: An Informal Economy Perspective* (New Delhi: NCEUS, 2009), 134. Available at http://nceuis.nic.in/The_Challenge_of_Employment_in_India.pdf.

47. Sara Sidner, "Odd Jobs Run India's Economy," CNN.com, October 14, 2009, http://edition.cnn.com/2009/BUSINESS/10/14/india.informal.economy/index.html.

48. The International Cooperative Alliance reports 800 million cooperative members worldwide. Considering that the combined population of the European Union and the United States is less than this number, and assuming that, at the very most, half of the people in these two regions own corporate shares, one can assume as many as several hundred additional million shareholders in the rest of the world and still arrive at less than 800 million. "Statistical Information on the Co-operative Movement," International Co-operative Alliance, www.ica.coop/members/member-stats.html.

49. "Statistical Information on the Co-operative Movement," International Co-operative Alliance.

50. Katherine Kobe, *The Small Business Share of GDP, 1998–2004* (Small Business Research Summary 299, Small Business Administration, Office of Advocacy, April 2007), 11. Available at www.smallbusinessnotes.com/pdf/rs299tot.pdf.

51. "Economy, Jobs Trump All Other Policy Priorities in 2009," Pew Research Center for People and the Press, survey report, January 22, 2009, http://people-press.org/report/485/economy-top-policy-priority.

52. "Most Ready for 'Green' Sacrifices," BBC News, November 7, 2007, http://news.bbc.co.uk/2/hi/7075759.stm.

53. Jon A. Krosnick, "The Climate Majority," *New York Times*, June 8, 2010, www.nytimes.com/2010/06/09/opinion/09krosnick.html?pagewanted=1&hp.

54. "Energy Update: Support for Renewable Energy Resources Reaches Highest Level Yet," Rasmussen Reports, January 11, 2011, www.rasmussenreports.com/public _content/politics/current_events/environment_energy/energy_update.

55. John Halpin and Karl Agne, *State of American Political Ideology, 2009: A National Study of Political Values and Beliefs* (Washington, DC: Center for American Progress, 2009), 10. www.glaserfoundation.org/program_areas/pdf/State%20of%20American %20Political%20Ideology%20-%20March%202009.pdf.

56. "Business Divides: Big Companies Have Too Much Power and Influence in DC, Small Business Has Too Little," Harris Polls, Harris Interactive, April 1, 2010, www .harrisinteractive.com/NewsRoom/HarrisPolls/tabid/447/ctl/ReadCustom%20 Default/mid/1508/ArticleId/115/Default.aspx.

57. Barack Obama, "Remarks by the President at the Acceptance of the Nobel Peace Prize" (speech, Oslo City Hall, Oslo, Norway, December 10, 2009), www.whitehouse .gov/the-press-office/remarks-president-acceptance-nobel-peace-prize.

58. Citigroup Research, "Plutonomy: Buying Luxury, Explaining Global Imbalance," Industry Note, Equity Strategy, October 16, 2005, 1–2. www.scribd.com/doc/6674234/ Citigroup-Oct-16-2005-Plutonomy-Report-Part-1. "Wealth" refers to net worth, includ- ing housing equity.

59. "March 2010 Trust in Government Survey," Pew Research Center for the People and the Press, survey report, March 21, 2010, http://people-press.org/reports /questionnaires/606.pdf, 105–106.

60. House Bill 4214, as passed by the Michigan Senate, March 9, 2011, www .legislature.mi.gov/documents/2011-2012/billengrossed/House/pdf/2011-HEBS -4214.pdf, 22.

61. Andy Kroll, "Behind Michigan's 'Financial Martial Law': Corporations and Right-Wing Billionaires," *Mother Jones*, March 23, 2011, 1. http://motherjones .com/politics/2011/03/michigan-snyder-mackinac-center.

62. David Horowitz, *The Art of Political War for Tea Parties* (Sherman Oaks, CA: David Horowitz Freedom Center, 2010), 10. www.discoverthenetworks.org/Articles/Tea %20party.pdf.

63. "March 2010 Trust in Government Survey," Pew Research Center for the People and the Press, survey report, March 21, 2010, 105–106.

64. Paul Steinhauser, "CNN Poll: Majority Think Government Poses Threat to Citi- zen's Rights," CNN Politics, February 26, 2010, http://articles.cnn.com/2010-02 -26/politics/citizens.rights.poll_1_national-poll-cnn-poll-survey?_s=PM:POLITICS.

65. Jonathon Porritt, *Living Within Our Means: Avoiding the Ultimate Recession* (London: Forum for the Future, 2009), 27. www.forumforthefuture.org/files/Living_within _our_means_sml.pdf. John Vidal, "Jonathon Porritt: 'I'm Not Going to Run Away and Be a Hippy,'" *Guardian*, July 25, 2009, www.guardian.co.uk/theguardian/2009/jul/25/ jonathan-porritt.

66. Hans-Peter Dürr, personal communication with author, May 2008.

67. Jason Palmer, "Brain Works More Like Internet Than 'Top Down' Company," BBC News, August 10, 2010, www.bbc.co.uk/news/science-environment-10925841?print=true.

68. Denis Noble, *The Music of Life: Biology Beyond Genes* (Oxford: Oxford University Press, 2006), 53.

69. E. O. Wilson, "Sustaining Life," *Resurgence* 261 (July–August 2010): 11.

70. Howard Schneider, "Greenspan: Regret over This 'Credit Tsunami,'" *Star Tribune*, October 23, 2008, www.startribune.com/business/33196929.html.

71. Peter S. Goodman, "The Reckoning: Taking Hard New Look at a Greenspan Legacy," *New York Times*, October 8, 2008, www.nytimes.com/2008/10/09/business/ economy/09greenspan.html?_r=1&dbk=&pagewanted=print.

## Thought Trap 1

1. Herman E. Daly, *Beyond Growth: The Economics of Sustainable Development* (Boston: Beacon Press, 1996), 31.

2. Erik Assadourian, "The Rise and Fall of Consumer Cultures," in *State of the World 2010*, ed. Worldwatch Institute (New York: W. W. Norton, 2010), 15.

3. How many pounds of grain and soy are consumed by one American steer to get one pound of edible meat? (a) The total forage (hay, silage, grass) consumed is 12,000 pounds (10,000 pre-feedlot and 2,000 in feedlot), and the total grain and soy-type concentrate consumed is about 2,850 pounds (300 pounds grain and 50 pounds soy before feedlot, plus 2,200 pounds grain and 300 pounds soy in feedlot). Therefore, the actual percentage of total feed units from grain and soy is about 25 percent. (b) But experts estimate that the grain and soy contribute more to weight gain (therefore to ultimate meat produced) than their actual proportion in the diet. They estimate that grain and soy contribute about 40 percent of weight put on over the life of the steer. (c) To estimate what percentage of edible meat is due to the grain and soy consumed, multiply that 40 percent (weight gain due to corn and soy) times the edible meat produced at slaughter, 432 pounds: 0.4 x 432 = 172.8 pounds of edible portion contributed by grain and soy. (Those who state a 7:1 ratio use the entire 432 pounds edible meat in their computation.) (d) To determine how many pounds of grain and soy it takes to get this

172.8 pounds of edible meat, divide total grain and soy consumed, 2,850 pounds, by 172.8 pounds of edible meat: 2,850 ÷ 172.8 = 16 to 17 pounds. (I have taken the lower figure, since the amount of grain and soy being fed may be going down a small amount.) These estimates are based on several consultations with the United States Department of Agriculture (USDA) Economic Research Service and the USDA Agricultural Research Service, Northeastern Division, plus newspaper reports of actual grain and soy being fed to cattle.

4. Robert Ayres and A. V. Kneese, "Externalities: Economics & Thermodynamics," in *Economy and Ecology: Towards Sustainable Development*, ed. Franco Archibugi and Peter Nijkamp (New York: Springer, 1989), 92.

5. US Environmental Protection Agency (EPA), *Municipal Solid Waste in the United States 2009: Facts and Figures* (Washington, DC: EPA, 2010), 171. www.epa.gov/osw/ nonhaz/municipal/pubs/msw2009rpt.pdf.

6. Lawrence Livermore National Laboratory and US Department of Energy, "Estimated U.S. Energy Use in 2009: ~94.6 Quads," Lawrence Livermore National Laboratory, August 2010, https://flowcharts.llnl.gov/content/energy/energy_archive/energy _flow_2009/LLNL_US_Energy_Flow_2009.png.

7. "America's Anemic '13 Percent Economy': Experts Warn U.S. Risks Long-Term Growth by Focusing on New Energy at Expense of More Energy Efficiency," American Council for an Energy-Efficient Economy, press release, April 28, 2010, www.aceee.org/ press/2010/04/americas-anemic-13-percent-economy-experts-warn-us-risks.

8. Janet L. Sawin and William R. Moomaw, *Renewable Revolution: Low-Carbon Energy by 2030* (Washington, DC: Worldwatch Institute, 2009), 9. www.worldwatch .org/files/pdf/Renewable%20Revolution.pdf.

9. "America's Anemic '13 Percent Economy,'" American Council for an Energy-Efficient Economy.

10. David Pimentel, Laura Westra, and Reed Noss, eds., *Ecological Integrity: Integrating Environment, Conservation and Health* (Washington, DC: Island Press, 2000), 129.

11. David Pimentel and Marcia Pimentel, "Sustainability of Meat-Based and Plant-Based Diets and the Environment," *American Journal of Clinical Nutrition* 78, no. 3 (2003): 662S. www.ajcn.org/content/78/3/660S.full.pdf+html.

12. Rex C. Buchanan, Robert R. Buddemeier, and B. Brownie Wilson, "The High Plains Aquifer," *Kansas Geological Survey: Public Information Circular* 18 (September 2001), www.kgs.ku.edu/Publications/pic18/PIC18R.pdf (revised December 2009).

13. Michelle Allsopp et al., "Oceans in Peril—Protecting Marine Biodiversity," *Worldwatch Report* 174 (September 2007): 6.

14. Deborah Zabarenko, "One-Third of World Fish Catch Used for Animal Feed," Reuters, October 29, 2008, www.reuters.com/article/2008/10/29/us-fish-food-idUSTRE49S0XH20081029.

15. Jeff Harrison, "Study: Nation Wastes Nearly Half Its Food," *UA News*, November 18, 2004, http://64.233.169.104/search?q=cache:6W_CSv7_kU0J:uanews.org/node/10448 +food+waste+arizona&hl=en&ct=clnk&cd=2&gl=us; see also Timothy W. Jones, "Using Contemporary Archaeology and Applied Anthropology to Understand Food Loss in the American Food System," Bureau of Applied Research in Anthropology, University of Arizona, www.ce.cmu.edu/~gdrg/readings/2006/12/19/Jones_UsingContemporary ArchaeologyAndAppliedAnthropologyToUnderstandFoodLossInAmericanFoodSystem.pdf.

16. Kevin D. Hall et al., "The Progressive Increase of Food Waste in America and Its Environmental Impact," *PLoS ONE* 4, no. 11 (November 25, 2009): 1. www.plosone.org/article/info%3Adoi%2F10.1371%2Fjournal.pone.0007940.

17. *Livestock's Long Shadow* (Rome: FAO, 2006), 43; feed calculated from Cereal Market Summary, *Food Outlook: Global Market Analysis* (Rome: FAO, May 2012), 1.

18. Colin Tudge, *Economic Renaissance: Holistic Economics for the 21st Century* (Devon, UK: Schumacher College/Green Books, 2008), 11.

19. Herman E. Daly and Joshua Farley, *Ecological Economics: Principles and Applications* (Washington, DC: Island Press, 2004), 4–5.

20. Herman E. Daly, "From a Failed-Growth Economy to a Steady-State Economy," *Solutions* 1, no. 2 (March–April 2010), 37. www.thesolutionsjournal.com/node/556.

21. Ahmed Djoghlaf, "Message from Mr. Ahmed Djoghlaf, Executive Secretary, on the Occasion of the International Day for Biological Diversity," United Nations Environment Program, May 22, 2007, http://www.cbd.int/doc/speech/2007/sp-2007-05-22-es-en.pdf.

22. "Experts Say That Attention to Ecosystem Services Is Needed to Achieve Global Development Goals," Millennium Ecosystem Assessment, press release, March 30, 2005, www.maweb.org/en/article.aspx?id=58.

23. David R. Montgomery, "Soil Erosion and Agriculture Sustainability," *Proceedings of the National Academy of Sciences* 104, no. 33 (2007): 13268–13272. www.pnas.org/content/104/33/13268.full?sid=7c7b19ba-8d41–4b3c-b80d-48f14333d595. Buchanan, Buddemeier, and Wilson, "The High Plains Aquifer"; "Key Facts About Air Pollution," American Lung Association, www.lungusa.org/assets/documents/key_air.pdf (accessed February 12, 2011); Robert S. Hoffman et al., *Goldfrank's Manual of Toxicologic Emergencies*, 8th ed. (New York: McGraw Hill, 2007), 1043.

24. Paul R. Epstein et al., "Full Cost Accounting for the Life Cycle of Coal," in "Ecological Economics Reviews," Robert Costanza, Karin Limburg and Ida Kubiszewski, eds., *Annals of the New York Academy of Sciences* 1219 (February 2011): 73–98. Available at http://onlinelibrary.wiley.com/doi/10.1111/j.1749–6632.2010.05890.x/full.

25. United Nations Environment Program (UNEP) Financial Initiative and PRI, "Universal Ownership: Why Externalities Matter to Institutional Investors," UNEP Financial Initiative, October 2010, www.unepfi.org/fileadmin/documents/universal_ownership .pdf, 6; Estimates are based on research by the London-based firm Trucost (www .trucost.com); Office of Management and Budget, *Budget of the U.S. Government: Fiscal Year 2011* (Washington, DC: GPO, 2010), 151, http://www.gpoaccess.gov/usbudget/ fy11/pdf/budget.pdf.

26. Jonathan Haughton and Shahidur Khandker, *Handbook on Poverty and Inequality* (Washington DC: The International Bank for Reconstruction and Development/ World Bank, 2009), 185, http://siteresources.worldbank.org/INTPA/Resources/429966 –1259774805724/Poverty_Inequality_Handbook_Ch10.pdf. See also "Concerning 400 Wealthiest: The Forbes 400 Special Report," *Forbes*, September 17, 2008, www.forbes .com/2008/09/16/forbes-400-billionaires-lists-400list08_cx_mn_0917richamericans _land.html. This article estimates that the four hundred wealthiest people in America have a combined net worth of $1.57 trillion. See also James Davies et al., "The Global Distribution of Household Wealth," *WIDER Angle* (Canada) 2 (December 1, 2006): 4–7. Global household wealth (defined as the value of physical and financial assets minus debts) in 2000 was valued at $125 trillion, and the poorest 50 percent of the world's adults owned barely 1 percent of global wealth, which would equal $1.25 trillion.

27. Robert B. Reich, *Supercapitalism: The Transformation of Business, Democracy, and Everyday Life* (New York: Knopf, 2007), 113. The family of Walmart founder Sam Walton has a combined fortune estimated to be about $90 billion. The combined wealth of the bottom 40 percent of the US population, comprising some 120 million people, in the same year, 2005, was estimated to be around $95 billion.

28. David Cay Johnston, *Free Lunch: How the Wealthiest Americans Enrich Themselves at Government Expense (and Stick You with the Bill)* (London: Portfolio Publishing, 2008), 275. Johnston is a Pulitzer Prize–winning journalist. In 2010, in private communication, he updated the estimate from a drop of $4,000 to $2,000 between 1973 and 2008.

29. Center for Responsive Politics, "Lobbying Spending Database—Koch Industries, 2008," OpenSecrets.org, www.opensecrets.org/lobby/clientsum.php?year=2008&lname =Koch+Industries&id= (accessed February 16, 2011). See also Jane Mayer, "Covert Operations: The Billionaire Koch Brothers Who Are Raging Against Obama," *New Yorker*,

August 30, 2010, www.newyorker.com/reporting/2010/08/30/100830fa_fact_mayer #ixzz1GgSAe397; "Toxic 100 Air Polluters Index," Political Economy Research Institute, March 21, 2010, www.peri.umass.edu/toxic_index; Pew Research Center for People and the Press, "Little Change in Opinions About Global Warming: Increasing Partisan Divide on Energy Policies," October 27, 2010, http://people-press.org/report/669/.

30. David D. Kirkpatrick and Charlie Savage, "Companies That Got Bailout Money Keep Lobbying," New York Times, January 23, 2009, www.nytimes.com/2009/01/24/business/24lobby.html.

31. Louise Story, "A Secret Banking Elite Rules Trading in Derivatives," New York Times, December 11, 2010, www.nytimes.com/2010/12/12/business/12advantage.html; Gretchen Morgenson, "Strong Enough for Tough Stains?" New York Times, June 25, 2010, www.nytimes.com/2010/06/27/business/27gret.html?ref=gretchen_morgenson;David Goldman, "Will the Reform Bill Prevent the Next Crisis?" CNN Money, May 24, 2010, http://money.cnn.com/2010/05/24/news/economy/preventing_next_crisis/index.htm.

32. Neil M. Barofsky, "Where the Bailout Went Wrong," New York Times, March 30, 2011, www.nytimes.com/2011/03/30/opinion/30barofsky.html.

33. Center for Responsive Politics, "Open Secrets Database," OpenSecrets.org, www.opensecrets.org/lobby/index.php. In 2010, there were 12,982 lobbyists. There are 535 members of Congress.

34. "Cutting Out the Middleman," New York Times, April 3, 2011.

35. President Franklin Delano Roosevelt, speech to joint session of Congress, April 29, 1938.

36. Doug Henwood (editor, Left Business Observer), personal communication with author, April 21, 2011. Calculations based on "U.S. National Economic Accounts," Bureau of Economic Analysis, www.bea.gov/national/nipaweb/TableView.asp?SelectedTable =87&Freq=Qtr&FirstYear=2008&LastYear=2010 (accessed April 28, 2011).

37. Jim Morris and M. B. Pell, "Renegade Refiner: OSHA Says BP Has 'Systemic Safety Problem,'" Center for Public Integrity: Investigations, May 16, 2010, www .publicintegrity.org/articles/entry/2085.

38. David Barstow et al., "Regulators Failed to Address Risks in Oil Rig Fail-Safe Device," New York Times, June 20, 2010, www.nytimes.com/2010/06/21/us/21blowout .html?pagewanted=1&_r=1.

39. Frances Beinecke, member of the National Commission on the Deepwater Horizon Oil Spill and Offshore Drilling, "Oil Spill Commission's Recommendations: Who's Started Adopting Them, Who Hasn't," Switchboard, March 30, 2011, http://switchboard .nrdc.org/blogs/fbeinecke/oil_spill_commissions_recommen.html; David Newman,

"BP Oil Disaster at One Year: Making the Most of the Natural Resource Damage Assessment," April 14, 2011, http://switchboard.nrdc.org/blogs/dnewman/one_year_update_on_deepwater_h.html; John M. Broder and Clifford Krauss, "Tight Regulation of Offshore Rigs Remains Elusive," *New York Times*, April 17, 2011, A1.

40. Jack Cloherty et al., "BP's Dismal Safety Record," ABC News, May 27, 2010, http://abcnews.go.com/WN/bps-dismal-safety-record/story?id=10763042&page=1; for profits, see John W. Schoen, "BP Faces Huge Tab, Has Deep Pockets," MSNBC, June 15, 2010, www.msnbc.msn.com/id/37689703/ns/business-world_business.

41. "Chemical Controls," *Scientific American*, April 2010, 1. www.scientificamerican.com/article.cfm?id=chemical-controls.

42. Michio Kaku and Jennifer Trainer, eds., *Nuclear Power: Both Sides—The Best Arguments For and Against the Most Controversial Technology* (New York: W. W. Norton, 1983), 113.

43. Amy Goodman, "Don't Drink the Nuclear Kool-Aid," Democracy Now!, July 17, 2008, www.democracynow.org/blog/2008/7/17/amy_goodmans_new_column_dont_drink_the_nuclear_kool_aid.

44. Center for Responsive Politics, "Lobbying Database: General Electric," OpenSecrets.org, www.opensecrets.org/lobby/clientsum.php?lname=General+Electric&year=2010 (accessed on April 21, 2011).

45. John McCormick, "Nuclear Illinois Helped Shape Obama View on Energy Dealings with Exelon," Bloomberg, March 23, 2011, www.bloomberg.com/news/2011-03-23/nuclear-illinois-helped-shape-obama-view-on-energy-in-dealings-with-exelon.html.

46. Oliver Tickell, "Toxic Link: The WHO and the IAEA," Guardian.co.uk, May 28, 2009, www.guardian.co.uk/commentisfree/2009/may/28/who-nuclear-power-chernobyl.

47. Helen Caldicott, "How Nuclear Apologists Mislead the World over Radiation," *Guardian*, April 11, 2011, www.guardian.co.uk/environment/2011/apr/11/nuclear-apologists-radiation. See also Alexey V. Yablokov et al., "Consequences of the Catastrophe for People and the Environment," *New York Academy of Sciences* 1181 (December 2009): 210; Chernobyl Forum, *Chernobyl's Legacy: Health, Environmental and Socio-Economic Impacts* (Vienna: IAEA, 2005), 14–15, http://www.iaea.org/Publications/Booklets/Chernobyl/chernobyl.pdf.

48. *Annual Energy Outlook 2011* (Washington, DC: EIA, 2010), 132, http://www.eia.gov/forecasts/aeo/pdf/0383%282011%29.pdf.

49. "Wind Generation vs. Capacity," U.S. Energy Information Administration, January 2011, www.eia.doe.gov/cneaf/solar.renewables/page/wind/wind.html (accessed on April 20, 2011).

50. Mark Schwartz et al., *Assessment of Offshore Wind Energy Resources for the United States* (Golden, CO: National Renewable Energy Laboratory, 2010), 4. www.nrel .gov/docs/fy10osti/45889.pdf. *Electric Power Industry 2009: Year in Review* (Washington, DC: EIA, 2009), www.eia.doe.gov/cneaf/electricity/epa/epa_sum.html.

51. "A Resurgence of Nuclear Power Poses Significant Challenges," Union of Concerned Scientists, May 2009, 1, www.ucsusa.org/assets/documents/nuclear_power/ nuclear-resurgence.pdf.

52. Union of Concerned Scientists (UCS), *Nuclear Power: Still Not Viable Without Subsidies* (Cambridge, MA: UCS Publications, 2011) 1, 3. www.ucsusa.org/nuclear _power/nuclear_power_and_global_warming/nuclear-power-subsidies-report.html.

53. "A Resurgence of Nuclear Power," 8-9.

54. "A Resurgence of Nuclear Power," 1.

55. US Government Accountability Office (GAO), "Federal Electricity Subsidies: Information on Research Funding, Tax Expenditures, and Other Activities That Support Electricity Production," GAO-08-102, October 2007, www.gao.gov/new.items/d08102.pdf, 2–3. Time period cited is in federal "fiscal years."

56. Nick Robins, Robert Clover, and Charanjit Singh, "A Climate for Recovery: The Color of Stimulus Goes Green," HSBC Global Research–Climate Change, February 25, 2009, http://globaldashboard.org/wp-content/uploads/2009/HSBC_Green_New_Deal.pdf, 2.

57. Daly, "From a Failed-Growth Economy to a Steady-State Economy," 38–39.

58. Adam Smith, *The Wealth of Nations* (New York: Random House, 1937), V.ii.2.777. References are to part, section, chapter, and paragraph.

59. Peter Lawrence, "Interface Chairman Ray C. Anderson on Sustainable Design," *@issue Journal* 13, no. 1 (2009), www.cdf.org/issue_journal/interfaces_chairman _ray_c._anderson_on_sustainable_design-2.html.

60. Janet Roberts, "Interface's Back-to-Nature Carpet Creates Big Profits," World Inquiry, Case Reserve University, 2004, http://worldinquiry.case.edu/feature_interface _000.cfm.

61. "InterfaceFLOR Takes the Lead with Europe's First Environmental Product Declaration (EPD) for Carpet Tiles," Interface Europe, news release, March 23, 2010, www.interfaceflor.eu/internet/web.nsf/webpages/571520107_EN.html; "Interface Flooring Systems Uses Landfill Emissions to Power Plant," Interface Flooring, press release, August 20, 2003, www.interfaceflooring.com/what/August20PressRelease.html; "Global Ecometrics: Greenhouse Gas Emissions," Interface, www.interfaceglobal .com/getdoc/7e96b54e-ad49-4eff-9877-38a55df0396d/Global-EcoMetrics.aspx (accessed March 15, 2011); Erin Meezan, "Interface Reduces Water Use 80% per Unit

Since 1996," Natural Step, news release, June 29, 2010, www.naturalstep.org/sv/usa/ interface-reduces-water-use-80-unit-1996.

62. Interview with Ray Anderson by Shelagh McNally, "Green Living Interview: Ray Anderson," *Green Living*, June 2008, http://www.greenlivingonline.com/article/green -living-interview-ray-anderson.

63. Lawrence, "Interface Chairman Ray C. Anderson on Sustainable Design."

64. National Crime Records Bureau, *Accidental Deaths & Suicides in India, 2006* (New Delhi: Ministry of Home Affairs, 2007), tables 2.10–2.11. http://ncrb.nic.in/adsi/ data/ADSI2006/home.htm.

65. Genetic Resources Action International (GRAIN), "Saying 'No' to Chemical Farming in India," *Seedling* (July 2008): 27–29. www.grain.org/seedling/?id=559.

66. GRAIN, "Saying 'No' to Chemical Farming in India." Vijaysekar Kalavadonda et al., *Ecologically Sound, Economically Viable: Community Managed Sustainable Agriculture in Andhra Pradesh, India* (Washington, DC: World Bank, 2009), 26. See also Savvy Soumya Misra, "Made It," *Down to Earth: Science and Environment Online* 20, no. 16 (January 1–15, 2009), www.downtoearth.org.in/cover.asp?foldername=20090115 &filename=news&sid=10&sec_id=9.

67. Kalavadonda et al., *Ecologically Sound, Economically Viable*, 9.

68. P. V. Satheesh, director, Deccan Development Society, e-mail message to author.

69. Satheesh, e-mail to author.

70. DDS Community Media Trust, P. V. Satheesh and Michel Pimbert, *Affirming Life and Diversity: Rural Images and Voices on Food Sovereignty in South India* (London: International Institute for Environment and Development, DDS, 2008), book/report and DVD-ROM.

71. "Navdanya's Organizational Overview," Navdanya.org, http://navdanya.org/ about-us/organization (accessed January 31, 2011).

72. For information, see the Business Alliance for Local Living Economies at www.livingeconomies.org.

73. "The Economic Impact of Locally Owned Businesses vs. Chains: A Case Study in Midcoast Maine," Institute for Local Self-Reliance, September 2003, www.newrules.org/retail/midcoaststudy.pdf.

74. Timothy Krantz, "Waste-Processing Technologies Transform the Town of Guss-ing," Institution of Engineering and Technology, August 2, 2010, http://kn.theiet.org/ magazine/issues/1012/waste-processing-1012.cfm.

75. Personal communication, David J. Thompson, March 2011; "Statistical Informa-tion on the Co-operative Movement," International Co-operative Alliance, January 31, 2007, www.ica.coop/members/member-stats.html.

76. David J. Thompson, "Clustering Coop Development, *Cooperative Grocer*, November 16, 2004, www.clcr.org/publications/other/emilia%20romagna%20by%20david%20thompson%20110604.pdf. On the number of worker-owners per enterprise, see Davide Pieri, official of Confcooperative, Bologna, Italy, personal communication with author, March 2011.

77. Gar Alperowitz, Ted Howard, and Thad Williamson, "The Cleveland Model," *Nation*, March 1, 2010, www.thenation.com/article/cleveland-model?page=0,0.

78. "Small Town Says, This Is Our Store," *Free Enterprise*, January 2009, www.uschambermagazine.com/article/small-town-says-this-is-our-store; Michelle Nijhuis, "For Sale, by Owners," *Smithsonian,* October 2004, www.smithsonianmag.com/people-places/For_Sale_By_Owners.html.

79. Rural Community Shops, "About Us," Plunkett Foundation, www.plunkett.co.uk/whatwedo/rcs/about.cfm (accessed September 10, 2010).

80. Christian Gelleri, "Chiemgauer Regiomoney: Theory and Practice of a Local Currency," *International Journal of Community Currency Research* 13 (2009): 61–75. www.uea.ac.uk/env/ijccr/pdfs/IJCCRvol13%282009%29pp61–75Gelleri.pdf; Judith D. Schwartz, "Buying Local: How It Boosts the Economy," *Time*, June 11, 2009, www.time.com/time/business/article/0,8599,1903632,00.html; Herbert Girardet and Miguel Mendonça, *A Renewable World: A Report for the Future World Council* (Totnes, UK: Green Books Ltd, 2009), 195–197; interviews with participants in Germany, 2009.

81. Judith D. Schwartz, "Dollars with Good Sense," *Yes! Magazine* 50 (summer 2009): 31.

82. "GNI per Capita, PPP," World Development Indicators Data, http://data.worldbank.org/indicator/NY.GNP.PCAP.PP.CD (accessed April 1, 2011).

83. Luis Rosero-Bixby, "Costa Rica Saves Infants' Lives," *World Health Forum* 9 (1988): 442, 443. http://whqlibdoc.who.int/whf/1988/vol9-no3/WHF_1988_9%283%29_p439–443.pdf.

84. Central Intelligence Agency, "Country Comparison: Life Expectancy at Birth," The World Factbook, https://www.cia.gov/library/publications/the-world-factbook/rankorder/2102rank.html?countryCode=&rankAnchorRow=# (accessed March 8, 2011).

85. Katia Karousakis, *Incentives to Reduce GHG Emissions From Deforestation: Lessons Learned From Costa Rica and Mexico* (Organisation for Economic Co-Operation and Development and International Energy Agency, 2007), 16–23, http://www.oecd.org/dataoecd/55/54/38523758.pdf.

86. Thomas L. Friedman, "(No) Drill, Baby, Drill," *New York Times*, April 12, 2009, www.nytimes.com/2009/04/12/opinion/12friedman.html.

87. Saamah Abdallahet et al., *The (Un)Happy Planet Index 2.0: Why the Good Life Doesn't Have to Cost the Earth* (London: New Economics Foundation, May 2009), 22, 28, 61.

88. P. Sainath, "Of Luxury Cars and Lowly Tractors," *Hindu*, December 27, 2010, www.thehindu.com/opinion/columns/sainath/article995828.ece#.

89. John Talberth, Clifford Cobb, and Noah Slattery, *The Genuine Progress Indicator 2006: A Tool for Sustainable Development* (Oakland, CA: Redefining Progress, 2006), 17.

90. Talberth et al., *The Genuine Progress Indicator*, 19–22.

91. Helen Kersley and Eilís Lawlor, *Grounded: A New Approach to Evaluating Runway 3* (London: New Economics Foundation, 2010), 15. www.neweconomics.org/sites/neweconomics.org/files/Grounded_0.pdf.

92. "Continued Public Support for Going 'Beyond GDP': Global Poll," GlobeScan, news release, January 21, 2011, www.globescan.com/news_archives/ethicalmarkets2011/Beyond_GDP_2011.pdf.

93. Jon Gertner, "The Rise and Fall of the GDP," *New York Times*, May 10, 2010, www.nytimes.com/2010/05/16/magazine/16GDP-t.html?pagewanted=all; *Vital Signs 2011* (Washington, DC: Worldwatch Institute, 2011), 76.

94. "Governor Martin O'Malley Launches Genuine Progress Indicator," Office of Governor Martin O'Malley, news release, February 3, 2010, www.gov.state.md.us/pressreleases/100203.asp. See also "Maryland's Genuine Progress Indicator," Maryland: Smart, Green & Growing, www.green.maryland.gov/mdgpi.

95. Neil Tickner, "Maryland to Measure Economic Conditions from Main St. View," UM Newsdesk, February 3, 2010, www.newsdesk.umd.edu/scitech/release.cfm?ArticleID=2076.

## Thought Trap 2

1. Eve Troeh, "New Diapers for Trendier Tushes," NPR Marketplace Report, June 30, 2010, http://marketplace.publicradio.org/display/web/2010/06/30/pm-new-diapers-for-trendier-tushes.

2. Robert Kennedy Jr., "A Bottle Bill That Will Rot Your Teeth," *New York Times*, May 27, 2009, www.nytimes.com/2009/05/28/opinion/28kennedy.html.

3. Mitchell Joachim, "The Re(f)use City and WALL·E," *Solutions* 1, no. 1 (January–February 2009): 22. www.thesolutionsjournal.com/perspective/2009-01-28-refuse-city-and-wall.

4. Richard Heinberg, "Look on the Bright Side," *Museletter* 206 (June 2, 2009), http://richardheinberg.com/206-look-on-the-bright-side.

5. Paul Ehrlich and Anne H. Ehrlich, *The Dominant Animal: Human Evolution Environment* (Washington, DC: Island Press, 2008), 217.

6. "Overconsuming Health," *Left Business Observer* 120 (August 2009), www .leftbusinessobserver.com/Consumption.html.

7. Office of Management and Budget, *Budget of the U.S. Government: Fiscal Year 2011* (Washington, DC: US Government Printing Office, 2010), 7. www.gpo.gov/ fdsys/pkg/BUDGET-2011-BUD/pdf/BUDGET-2011-BUD.pdf.

8. Tim Jackson, *Prosperity Without Growth: Economics for a Finite Planet* (London: Earthscan, 2009), 188.

9. John Naish, "We Have Enough," *Resurgence* 159 (March–April 2010): 17.

10. Xinyue Zhou, Kathleen D. Vohs, and Roy F. Baumeister, "The Symbolic Power of Money: Reminders of Money Alter Social Distress and Physical Pain," *Psychological Science* 20, no. 6 (2009): 700–706.

11. Tom Crompton and Tim Kasser, *Meeting Environmental Challenges: The Role of Human Identity* (London: WWF-UK, 2009), 18–22.

12. "No Growth in Global Adspend 2009," ZenithOptimedia, press release, December 8, 2008, www.vivaki.at/fileadmin/user_upload/_pdf/2008-12-08_PM _AEF_englisch.pdf.

13. Charles U. Larson, *Persuasion: Reception and Responsibility* (Boston: Wadsworth, Cengage Learning, 2009), 378.

14. US Census Bureau, *Current Population Survey, 2010: Annual Social and Economic Supplement* (Washington, DC: US Census Bureau, 2010), table HINC-06. www.census .gov/hhes/www/cpstables/032010/hhinc/new06_000.htm.

15. Citigroup Research, "Plutonomy: Buying Luxury, Explaining Global Imbalance," Industry Note, Equity Strategy, October 16, 2005, 1–2. Available at www.scribd.com/ doc/6674234/Citigroup-Oct-16-2005-Plutonomy-Report-Part-1. "Wealth" refers to net worth, including housing equity.

16. "Bill Gates No Longer World's Richest Man," *Forbes*, March 10, 2010, www.forbes.com/2010/03/09/worlds-richest-people-slim-gates-buffett-billionaires -2010-intro.html; World Bank, *2010 World Development Indicators* (Washington, DC: International Bank for Reconstruction and Development/World Bank, 2010), 137. http://data.worldbank.org/sites/default/files/wdi-final.pdf.

17. David Rosnick and Mark Weisbrot, "Are Shorter Work Hours Good for the Environment? A Comparison of U.S. and European Energy Consumption," Center for Economic and Policy Research, December 2006, 3, 7, www.cepr.net/documents/publications/ energy_2006_12.pdf; Richard Rogerson, "Structural Transformation and the Deterioration of European Labor Market Outcomes," *Journal of Political Economy* 116, no. 2 (2008): 235–259.

18. Interview with Lester Brown by Steve Curwood, "How Close Are We to the Edge," *Living on Earth*, National Public Radio, www.loe.org/shows/segments.htm ?programID=11-P13-00007&segmentID=6.

19. Christopher Delgado et al., *Livestock to 2020: The Next Food Revolution* (Food Agriculture, and the Environment Discussion Paper 28, International Food Policy Research Institute, Food and Agriculture Organization of the United Nations, International Livestock Research Institute, May 1999), table 6; for cereal use as feed, see table 15. http://dspace.ilri.org:8080/jspui/bitstream/10568/333/2/dp28.pdf.

20. International Fund for Agricultural Development (IFAD), *Rural Poverty Report 2011* (Rome: IFAD, 2010), 47. www.ifad.org/rpr2011/report/e/rpr2011.pdf. "Very poor" refers to those living on less than $1.25 per day. For more on the potential of smallholder ecological farming, see Thought Traps 1 and 3.

21. Christopher Solomon, "For Many Homeowners, Less Is So Much More," MSN Real Estate, http://realestate.msn.com//article.aspx?cp-documentid=13107878.

22. Alan E. Pisarski, *Commuting in America III* (Washington, DC: Transportation Research Board of the National Academies, 2006), 51, 101. http://onlinepubs.trb.org/onlinepubs/nchrp/ciaiii.pdf.

23. Federal Highway Administration, "Recent Trends in Congestion," in *Traffic Congestion and Reliability: Trends and Advanced Strategies for Congestion Mitigation* (Washington, DC: US Department of Transportation, 2005), www.ops.fhwa.dot.gov/congestion_report/chapter3.htm.

24. Rich Pirog, *Food, Fuel, and Freeways: An Iowa Perspective on How Far Food Travels, Fuel Usage and Greenhouse Gas Emissions* (Ames: Iowa State University, Leopold Center for Sustainable Agriculture, 2001), www.leopold.iastate.edu/pubs/staff/ppp/food_mil.pdf.

25. The definition is based on the following sources: "Passive Solar Design," Sustainable Sources, http://passivesolar.sustainablesources.com/#Define (accessed March 10, 2011); Yvonne Jeffrey, Liz Barclay, and Michael Grosvenor, "Regulating Your Home's Climate with Solar Techniques," Dummies.com, www.dummies.com/how-to/content/regulating-your-homes-climate-with-solar-technique.html (accessed March 10, 2011).

26. For an example of active solar in action, see "Hørsholm: Denmark's Most Climate Friendly Kindergarten," Sustainable Cities Best Practices Database, http://sustainablecities.dk/en/city-projects/cases/hoersholm-denmarks-most-climate-friendly-kindergarten.

27. Wendy Koch, "New Homes Being Built Smaller," *USA Today*, January 11, 2009, www.usatoday.com/money/economy/housing/2009-01-08-homesize_N.htm.

28. "The United States Summary Information," Farm Subsidies Data Base, http://farm.ewg.org/region?fips=00000&regname=UnitedStatesFarmSubsidySummary (accessed March 1, 2011).

29. Gary Paul Nabhan, *Coming Home to Eat: The Pleasures and Politics of Local Foods* (New York: W. W. Norton & Company, 2002), 72.

30. Rich Morin and Paul Taylor, "Luxury or Necessity? The Public Makes a U-Turn," Pew Research Center, April 23, 2009, http://pewresearch.org/pubs/1199/more-items -seen-as-luxury-not-necessity?src=prc-latest&proj=peoplepress.

31. Junko Edahiro, "Letter from Japan," *Resurgence* 262 (September–October 2010), 16.

32. Lindsay Hoshaw, "Afloat in the Ocean, Expanding Islands of Trash," *New York Times*, November 9, 2009, www.nytimes.com/2009/11/10/science/10patch.html?_r=3.

33. Charles Moore, "Out in the Pacific, Plastic Is Getting Drastic," Marine Litter, http://marine-litter.gpa.unep.org/documents/World's_largest_landfill.pdf. Moore is re-ferring to the area of the plastic soup in the central Pacific Ocean.

34. Rex C. Buchanan, Robert R. Buddemeier, and B. Brownie Wilson, "The High Plains Aquifer," *Kansas Geological Survey: Public Information Circular* 18 (September 2001), www.kgs.ku.edu/Publications/pic18/PIC18R.pdf (revised December 2009).

35. Bob Feldman, "War on the Earth," *Dollars and Sense* (March–April 2003), 24. www.dollarsandsense.org/archives/2003/0303maps.pdf.

36. Center for Responsive Politics, "Lobbying Database," OpenSecrets.org, www.opensecrets.org/lobby/index.php (accessed January 3, 2011).

37. *Who Killed the Electric Car?* directed by Chris Paine (Park City, UT: Sony Pictures Classics, 2006), DVD; Michael Taylor, "Owners Charged Up Over Electric Cars, but Man-ufacturers Have Pulled the Plug," *San Francisco Chronicle*, April 24, 2005, http://articles .sfgate.com/2005-04-24/news/17367829_1_electric-car-program-super-ultra-low -emission-vehicles-sulev.

38. Peter Gleick, *Bottled and Sold: The Story Behind Our Obsession with Bottled Water* (Washington, DC: Island Press, 2010), ix, x.

39. Jef I. Richards and Catherine M. Curran, "Oracles on 'Advertising': Searching for a Definition," *Journal of Advertising* 31, no. 2 (2002): 63.

40. World Bank, *World Development Indicators 2010* (Washington, DC: International Bank for Reconstruction and Development/World Bank, 2010), 94–96. http://data .worldbank.org/sites/default/files/wdi-final.pdf; "Income Share Held by Lowest 20%," World Development Indicators, http://data.worldbank.org/indicator/SI.DST.FRST.20 (accessed February 28, 2011).

41. Tim Kasser, *The High Price of Materialism* (Cambridge, MA: MIT Press, 2003), 33–53. "Cohens reported that children from families of low socioeconomic background also admired materialistic values more and placed a higher priority on 'being rich'" (33). "Ingelhart and colleagues . . . surveyed almost 50,000 people from 40 societies around the world. Parallel to the results concerning family socioeconomic status, people in poorer nations, who presumably experienced more insecurity, were typically more materialistic than those in wealthier nations" (35).

42. National Center for Chronic Disease Prevention and Health Promotion, Health Topics, Nutrition, Nutrition Fact Sheet. http://www.cdc.gov/healthyyouth/nutrition/facts.htm.

43. Health Code—"Setting Nutritional Standards for Restaurant Food Sold Accompanied by Toys or Other Youth Focused Incentive Items," San Francisco Ordinance No. 290–10 (passed November 2, 2010). See also "San Francisco Approves Ban Limits on Toys with Fast-Food Meals," Reuters, November 9, 2010, www.reuters.com/article/idUSTRE6A90DX20101110.

44. Larry Rohter, "Billboard Ban in São Paulo Angers Advertisers," *New York Times*, December 12, 2006, www.nytimes.com/2006/12/12/world/americas/12iht-brazil.html; interview with Vinicius Galvao by Bob Garfield, "Clearing the Air," *On the Media*, National Public Radio, April 20, 2007, www.onthemedia.org/transcripts/2007/04/20/04; "'Clean City': São Paulo Scrubbed of Outdoor Ads," WebUrbanist, http://weburbanist.com/2010/03/06/clean-city-sao-paulo-scrubbed-of-outdoor-ads (accessed February 3, 2011).

45. Paula Alvarado, "Buenos Aires to Remove 40 Thousand Billboards to Fight Visual Pollution," TreeHugger, August 12, 2008, www.treehugger.com/files/2008/08/buenos-aires-anti-visual-pollution-plan-remove-billboards.php.

46. Catherine Gudis, *Buyways: Billboards, Automobiles and the American Landscape* (New York: Routledge, 2004), 226.

47. Interview with Thomas Berry by Caroline Webb, *The Earth Community*, February 2006, www.earth-community.org/images/BerryIV_Subtitles.mov.

## Thought Trap 3

1. Jonathon Porritt, "Living Within Our Means," *Resurgence* 255 (July–August 2009): 10.

2. For information on Earth Hour 2008, visit www.earthhour2008.com. For later years, visit Earth Hour at www.earthhour.org.

3. Tim Kasser, *The High Price of Materialism* (Cambridge, MA: MIT Press, 2003), 29.

4. Ker Than, "Earth Hour 2010: Record 121 Countries to Go Dark," National Geographic Daily News, March 26, 2010, http://news.nationalgeographic.com/2010/03/100326-earth-hour-2010-record-landmarks.

5. The 5 percent estimate comes from Jeff Johnson, "The End of the Light Bulb," *Chemical & Engineering News* 85, no. 49 (2007): 46–51. http://pubs.acs.org/cen/government/85/8549gov1.html; the Edison quote about solar comes from James Newton, *Uncommon Friends: Life with Thomas Edison, Henry Ford, Harvey Firestone, Alexis Carrel & Charles Lindbergh* (Boston: Houghton Mifflin Harcourt, 1989), 31.

6. Christopher Flavin, "Building a Low-Carbon Economy," in *State of the World 2008*, ed. Worldwatch Institute (New York: W. W. Norton, 2008), 75. at www.worldwatch.org/files/pdf/SOW08_chapter_6.pdf.

7. Herman E. Daly and Joshua Farley, *Ecological Economics: Principles and Applications* (Washington, DC: Island Press, 2004), 10.

8. Bernard Mulligan and Mary Wildermuth, *The World of Energy: A Textbook for Physics 104* (Columbus: Ohio State University, Department of Physics, 2006), 27. www.physics.ohio-state.edu/104/textbook_pdf/per3.pdf.

9. Tim Jackson, *Prosperity Without Growth: Economics for a Finite Planet* (London: Earthscan, 2009), 119.

10. Gunter Pauli, *The Blue Economy: 10 Years, 100 Innovations, 100 Million Jobs* (Taos, NM: Paradigm Publications, 2010), 34.

11. William McDonough and Michael Braungart, *Cradle to Cradle: Remaking the Way We Make Things* (New York: North Point Press, 2002).

12. See the work of Gunter Pauli and ZERI at www.zeri.org.

13. "A Winning Combo: Coffee Waste and Mushrooms," *The Specialty Coffee Chronicle* 3 (May 2009): 8–9; Pauli, *The Blue Economy*, 88, 90, 240.

14. "Water Hyacinth Spawns Mushroom Enterprise," New Agriculturalist, May 1, 2003, www.new-ag.info/03-3/develop/dev04.html.

15. Paul Stamets, *Mycelium Running: How Mushrooms Can Help Save the World* (Berkeley, CA: Ten Speed Press, 2005).

16. Gunter Pauli, *The Blue Economy,* 269, *Vital Signs,* 83.

17. Elizabeth Rosenthal, "Using Waste, Swedish City Cuts Its Fossil Fuel Use," *New York Times*, December 11, 2010, www.nytimes.com/2010/12/11/science/earth/11fossil.html.

18. Elizabeth Rosenthal, "Europe Finds Clean Energy in Trash, but U.S. Lags," *New York Times*, April 12, 2010, www.nytimes.com/2010/04/13/science/earth/13trash.html.

19. Jim Witkin, "Skiing Your Way to 'Hedonistic Sustainability,'" *New York Times* Green (blog), February 16, 2011, http://green.blogs.nytimes.com/2011/02/16/skiing -your-way-to-hedonistic-sustainability/?partner=rss&emc=rss; "New Moguls of Clean Energy," *Living on Earth*, Public Radio International, February 4, 2011, www.loe.org/ shows/segments.htm?programID=11-P13–00005&segmentID=8.

20. P. Ozge Kaplan, Joseph Decarolis, and Susan Thorneloe, "Is It Better to Burn or Bury Waste for Clean Electricity Generation," *Environmental Science and Technology* 43, no. 6 (2009): 1711–1717. http://pubs.acs.org/doi/pdf/10.1021/es802395e.

21. "Municipal Solid Waste Factsheet" (Pub. No. CSS04–15), Center for Sustainable Systems, University of Michigan, 2009, http://css.snre.umich.edu/css_doc/CSS04 -15.pdf.

22. For US waste going to landfill, see "Wastes–Non-Hazardous Waste–Municipal Solid Waste," US Environmental Protection Agency, www.epa.gov/epawaste/nonhaz/ municipal/index.htm (last modified February 16, 2011). For New York City, see Rosenthal, "Europe Finds Clean Energy in Trash, but U.S. Lags"; for Germany and the Netherlands, see "40% of Municipal Waste Recycled or Composted in 2008," Eurostat, news release, March 19, 2010, http://epp.eurostat.ec.europa.eu/cache/ITY_PUBLIC/8 -19032010-AP/EN/8-19032010-AP-EN.PDF.

23. Barbara Praetorius et al., *Innovation for Sustainable Electricity Systems: Exploring the Dynamics of Energy Transitions* (Heidelberg, Germany: Physica-Verlag, 2008), 57–58.

24. *Technical and Economic Assessment of Combined Heat and Power Technologies for Commercial Customer Applications* (Palo Alto, CA, and Birmingham, AL: EPRI and Southern Company Services, 2003), 2-1. http://my.epri.com/portal/server.pt?space =CommunityPage&cached=true&parentname=ObjMgr&parentid=2&control=Set Community&CommunityID=404&RaiseDocID=000000000001007759&RaiseDocType =Abstract_id.

25. International Energy Association, *Cogeneration and District Energy: Sustainable Energy Technologies for Today and Tomorrow* (Paris: International Energy Association, 2009), 15. www.iea.org/files/CHPbrochure09.pdf; Danish Energy Agency, *Energi Statistics 2009* (Copenhagen: Danish Energy Agency, 2010), 54. http://ens.dk/en-US/Info/ FactsAndFigures/Energy_statistics_and_indicators/Annual%20Statistics/Documents/ Energi%20Statistics%202009.pdf.

26. Janine M. Benyus, *Biomimicry: Innovation Inspired by Nature* (New York: Harper Perennial, 1997), 6.

27. Sara J. Scherr and Sajal Sthapit, "Mitigating Climate Change Through Food and Land Use," *Worldwatch Report* 179 (2009): 8.

28. For total emissions, see Nicholas Stern, "Action and Ambition for a Global Deal in Copenhagen," United Nations Environment Program, December 6, 2009, www.unep.org/pdf/climatechange/actionandambitionforglobaldealincopenhagen .pdf. The 47 billion metric tons is an estimate for 2010 and is in the higher range of expert estimates. The 25 billion metric tons is derived from taking the more conservative estimate and applying to the current period the historical average percentage of greenhouse gases absorbed. Absorption calculated from R. A. Houghton, "Balancing the Global Carbon Budget," *Annual Review of Earth and Planetary Sciences* 35 (2007): 313–347.

29. This is calculated using data from William F. Ruddiman, "The Anthropogenic Greenhouse Era Began Thousands of Years Ago," *Climatic Change* 61, no. 3 (2003): 261–293; Thomas Boden, Gregg Marland, and Robert Andres, "Global, Regional and National Fossil-Fuel $CO_2$ Emissions," Carbon Dioxide Information Analysis Center, http://cdiac.ornl.gov/trends/emis/tre_glob.html (accessed February 21, 2011).

30. This is calculated using data from Boden, Marland, and Andres, "Global, Regional and National Fossil-Fuel $CO_2$ Emissions"; Richard Houghton, "Carbon Flux to the Atmosphere from Land-Use Changes: 1850–2005," in *TRENDS: A Compendium of Data on Global Change* (Oak Ridge, TN: Carbon Dioxide Information Analysis Center, 2008), http:// cdiac.ornl.gov/trends/landuse/houghton/houghton.html (accessed February 21, 2011).

31. Food and Agriculture Organization of the United Nations (FAO), *Global Forest Resources Assessment 2010* (Rome: FAO, 2010), xiii, 17. www.fao.org/docrep/013/ i1757e/i1757e.pdf.

32. *Common Sense on Climate Change: Practical Solutions to Global Warming* (Cambridge, MA: Union of Concerned Scientists, n.d.), 8.

33. Tim J. LaSalle and Paul Hepperly, *Regenerative Organic Farming: A Solution to Global Warming* (Kutztown, PA: Rodale Institute, 2008), 1. www.rodaleinstitute.org/files/ Rodale_Research_Paper-07_30_08.pdf.

34. LaSalle and Hepperly, *Regenerative Organic Farming*, 2.

35. Jerry D. Glover, Cindy M. Cox, and John P. Reganold, "Future Farming: A Return to Roots?" *Scientific American* (August 2007): 82–89.

36. Henning Steinfeld et al., *Livestock's Long Shadow: Environmental Issues and Options* (Rome: FAO, 2006), xxi. ftp://ftp.fao.org/docrep/fao/010/a0701e/a0701e.pdf.

37. Steinfeld et al., *Livestock's Long Shadow*, xxi.

38. Steinfeld et al., *Livestock's Long Shadow*, 12, 43. See also Anna Lappé, *Diet for a*

*Hot Planet: The Climate Crisis at the End of Your Fork and What You Can Do About It* (New York: Bloomsbury, 2010).

39. Allan Savory, "A Global Strategy for Addressing Global Climate Change," Savory Institute, 2008, www.savoryinstitute.com/storage/articles/A%20Global%20Strategy %20for%20Addressing%20Climate%20Change%202%20_1_.pdf, 13; "Carbon Storage," World Resources Institute, www.wri.org/publication/content/8272.

40. Steinfeld et al., *Livestock's Long Shadow*, 272; Savory, "A Global Strategy for Addressing Global Climate Change," 14.

41. Pete Smith et al., "Greenhouse Gas Mitigation in Agriculture," Philosophical Transactions of the Royal Society, B. 363, no. 1492 (February 2008), http://rstb .royalsocietypublishing.org/content/363/1492/789.full.pdf+html; for a higher estimate, see LaSalle and Hepperly, *Regenerative Organic Farming*, 1. "Even though climate and soil type affect sequestration capacities, these multiple research efforts verify that practical organic agriculture, if practiced on the planet's 3.5 billion tillable acres, could sequester nearly 40 percent of current $CO_2$ emissions."

42. United Nations Environment Program (UNEP), "Trees on Farms Key to Climate and Food-Secure Future," UNEP, July 24, 2009, www.unep.org/Documents.Multilingual/ Default.asp?DocumentID=593&ArticleID=6256&l=en; see also Festus K. Akinnifesi et al., "Fertiliser Trees for Sustainable Food Security in the Maize-Based Production Systems of East and Southern Africa: A Review," *Agronomy for Sustainable Development* 30, no. 3 (2010).

43. Adam Nossiter, "Famine Persists in Niger, but Denial Is Past," *New York Times*, May 3, 2010, www.nytimes.com/2010/05/04/world/africa/04niger.html.

44. Chris Reij et al., "Agroenvironmental Transformation in the Sahel" (Discussion Paper 00194, International Food Policy Research Institute, Washington, DC, November 2009), 2, 7, 19. www.ifpri.org/sites/default/files/publications/ifpridp00914.pdf.

45. Alex Perry, "Land of Hope," Time.com, December 13, 2010, www.time.com/ time/magazine/article/0,9171,2034377-2,00.html.

46. Mark Hertsgaard, "Regreening Africa," *The Nation*, November 19, 2009, www.thenation.com/article/regreening-africa?page=0,0.

47. Chris Reij, personal communication with author, June 30, 2010.

48. Reij et al., "Agroenvironmental Transformation in the Sahel," 33.

49. Chris Reij, "Investing in Trees to Mitigate Climate Change," in *State of the World 2011*, ed. Worldwatch Institute (New York: W. W. Norton, 2011), 88.

50. Frances Moore Lappé and Anna Lappé, *Hope's Edge: The Next Diet for a Small Planet* (New York: Tarcher, 2002).

51. C. Munster et al., "Trees on Farms: Tackling the Triple Challenge of Mitigation, Adaptation, and Food Security" (World Agroforestry Centre Policy Brief 07, World Agroforestry Center, Nairobi, 2009), 3.

52. Gert Jan Nabuurs et al., "Forestry," in *Climate Change 2007: Impacts, Adaptation and Vulnerability*, ed. Intergovernmental Panel on Climate Change (Cambridge: Cambridge University Press, 2007), 543. www.ipcc.ch/pdf/assessment-report/ar4/wg3/ar4-wg3-chapter9.pdf.

53. Paul Blackwell, Glen Riethmuller, and Mike Collins, "Biochar Application to Soil," in *Biochar for Environmental Management*, ed. Johannes Lehmann and Stephen Joseph (London: Earthscan, 2009); Joseph M. Kimetu et al., "Reversibility of Soil Productivity Decline with Organic Matter of Differing Quality Along a Degradation Gradient," *Ecosystems* 11 (2008): 726–739. www.css.cornell.edu/faculty/lehmann/publ/Ecosystems %2011,%20726–739,%202008%20Kimetu.pdf; "Cameroon Trial Data Show Strong Improvement in Maize Yields" (summary report, Biochar Fund, September 2009), www.biocharfund.org/index.php?option=com_content&task=view&id=54&Itemid=74.

54. For background, see the Biochar Fund at www.biocharfund.org; for the Congo initiative, see the Congo Basin Forest Fund at www.cbf-fund.org/index.php. See also "Innovative Biochar Project Wins Major Funding for Protection of Rainforests in Congo," Biochar Fund, press release, May 20, 2009, http://biocharfund.org/index.php?option=com_content&task=view&id=43.

55. "Nature Communications Article Shows 'True Colours' of Biochar Advocates: Groups Condemn Implied Land-Grab for Biochar," ETC Group, press release, August 30, 2010, www.etcgroup.org/en/node/5198.

56. *State of Food Insecurity in the World 2012*, Food and Agriculture Organization of the United Nations, (Rome: FAO, 2012), Figure A2.2, 55.

57. Joel K. Bourne, Jr., "The Global Food Crisis: The End of Plenty," *National Geographic*, June 2009, http://ngm.nationalgeographic.com/2009/06/cheap-food/bourne-text.

58. Bill McKibben, *Eaarth* (New York: Henry Holt and Company, 2010), 153

59. Robert Engelman, "An End to Population Growth," *Solutions Journal*, April, 2011, http://www.thesolutionsjournal.com/node/919; Population Growth Steady in Face of a Changing Climate, http://www.worldwatch.org/node/6262

60. Food and Agriculture Organization of the United Nations, FAOSTAT. For production increase, Production, http://faostat.fao.org/site/612/default.aspx#ancor; for calories, Food Balance Sheets, http://faostat.fao.org/site/368/default.aspx#ancor;

for "least developed," Production Indices, http://faostat.fao.org/site/612/default
.aspx#ancor.

61. Madeleine Bunting, "How Can We Feed the World and Still Save the Planet?"
*Guardian*, January 21, 2011.

62. Bunting, "How Can We Feed the World . . . "; ETC Group, *Who Will Feed Us? Questions for the Food and Climate Crisis*, 2009. www.etcgroup.org; Bill Vorley and the *UK Food Group, Food, Inc. Corporate Concentration from Farm to Consumer*, International Institute for Environment and Development, 2003.

63. Frederick Kaufman, "How Goldman Sachs Created the Food Crisis," *Foreign Policy*, April 27, 2011, http://www.foreignpolicy.com/articles/2011/04/27/how_goldman
_sachs_created_the_food_crisis?page=0,0.

64. Food Price Index, FAO http://www.fao.org/worldfoodsituation/wfs-home
/foodpricesindex/en/; Klaus Deininger and Derek Byerlee, *Rising Global Interest in Farmland* (Washington, DC: International Bank for Reconstruction and Development/
World Bank, 2011), xiv, xxxii, 55. http://siteresources.worldbank.org/INTARD/Resources
/ESW_Sept7_final_final.pdf

65. Lester Brown, "The Great Food Crisis of 2011," *Foreign Policy*, January 10, 2011,
www.foreignpolicy.com/articles/2011/01/10/the_great_food_crisis_of_201; John Baffes
and Tassos Haniotis, *Placing the 2006/08 Commodity Price Boom into Perspective*
(Washington, DC: World Bank, 2010), 11–12. http://blogs.worldbank.org/prospects
/placing-the-200608-commodity-price-boom-into-perspective.

66. For evidence of excellent harvests, "Production Indices," FAOSTAT).

67. For overviews of issues in this section, see Robin Broad and John Cavanagh, *Development Redefined* (London & Boulder: Paradigm, 2009; Walden Bello, *The Food Wars*
(London: Verso, 2009).

68. Rachel Cleetus, Steven Clemmer, and David Friedman, *Climate 2030: A National Blueprint for a Clean Energy Economy* (Cambridge, MA: Union of Concerned Scientists, 2009), 1, 4. www.ucsusa.org/global_warming/solutions/big_picture_solutions/climate
-2030-blueprint.html.

69. Rocky Mountain Institute, "Project Case Study: Empire State Building," RetroFit Depot, www.retrofitdepot.org/Content/Files/ESBCaseStudy.pdf; for superwindow efficiency, see David Wann, Center for Resource Management, *Deep Design: Pathways to a Livable Future* (Washington, DC: Island Press, 1996), 89.

70. Paul Hawken, Amory B. Lovins, and L. Hunter Lovins, *Natural Capitalism*, rev. ed.
(Boston: Little, Brown and Company, 2010), xiv.

71. Hawken, Lovins, and Lovins, *Natural Capitalism*, 115–119.

72. "GDP per Unit of Energy Use (Constant 2005 PPP $ per kg of Oil Equivalent)," World Bank Indicators, http://data.worldbank.org/indicator/EG.GDP.PUSE.KO.PP.KD.

73. Eric Beinhocker et al., "The Carbon Productivity Challenge: Curbing Climate Change and Sustaining Economic Growth," McKinsey Global Institute, June 2008, www.mckinsey.com/mgi/reports/pdfs/Carbon_Productivity/MGI_carbon_productivity_full_report.pdf, 1718.

74. Diana Farrell et al., "The Case for Investing in Energy Productivity," McKinsey Global Institute, February 2008, www.mckinsey.com/mgi/reports/pdfs/Investing_Energy_Productivity/Investing_Energy_Productivity.pdf, 22.

75. National Council for Science and the Environment, David E. Blockstein, and Leo Wiegman, *The Climate Solutions Consensus: What We Know and What to Do About It* (Washington, DC: Island Press, 2010), 132.

76. Paul R. Epstein et al., "Full Cost Accounting for the Life Cycle of Coal," *Annals of the New York Academy of Sciences* 1219 (February 2011): 73–98. http://onlinelibrary.wiley.com/doi/10.1111/j.1749-6632.2010.05890.x/full.

77. Eric A. Finkelstein, Ian C. Fiebelkorn, and Guijing Wang, "National Medical Spending Attributable to Overweight and Obesity: How Much, and Who's Paying?" *Health Affairs*, May 14, 2003, W3–224, http://content.healthaffairs.org/content/early/2003/05/14/hlthaff.w3.219.full.pdf.

78. "New Evidence Confirms the Nutritional Superiority of Plant-Based Organic Foods," Organic Center, press release, March 18, 2008, www.organic-center.org/news.pr.php?action=detail&pressrelease_id=22.

79. Interview with Marshall Herskovitz by Bruce Gellerman, *Living on Earth*, National Public Radio, March 11, 2011, www.loe.org/shows/segments.html?programID=11-P13-00010&segmentID=7.

80. Matt Krantz, "Exxon Quarterly Profit at $10.4B," *USA Today*, July 28, 2006, www.usatoday.com/money/companies/earnings/2006-07-27-exxon_x.htm. Chart updated to 2008: $25.3 billion in 2004; $45.2 billion in 2008. For seconds a year: $45,220,000,000 ÷ 31,536,000 = $1,434 per second.

81. Federal Ministry of Environment et al., *Niger Delta Natural Resource Damage Assessment and Restoration Project* (Gland, Switzerland: International Union for Conservation of Nature, 2006), 1. http://cmsdata.iucn.org/downloads/niger_delta_natural_resource_damage_assessment_and_restoration_project_recommendation.doc.

82. Tom O'Neill, "The Curse of Black Gold," *National Geographic*, February 2007, http://ngm.nationalgeographic.com/2007/02/nigerian-oil/oneill-text.

83. Energy Information Administration, "Countries: Overview," May 24, 2011, http://www.eia.doe.gov/country/country_energy_data.cfm?fips=NI; Elisha Bala-Gbogbo, "Nigeria's Oil Revenue Rose 46% to $59 Billion in 2010," *Bloomberg Businessweek,* April 14, 2011, http://www.businessweek.com/news/2011-04-14/nigeria-s-oil-revenue-rose-46-to-59-billion-in-2010.html; O'Neill, "The Curse of Black Gold."

84. "Post-War Expansion," Shell, www.shell.com/home/content/aboutshell/who_we_are/our_history/post_war_expansion (accessed April 6, 2011); Rebekah Kebede, "Shell Nigeria Case May Temper Big Oil Policies," Reuters, June 18, 2009, www.reuters.com/article/idUSTRE55H6A620090618.

85. Thomas L. Friedman, "The Green Revolution(s)," *New York Times,* June 24, 2009, www.nytimes.com/2009/06/24/opinion/24friedman.html.

86. "Frequently Asked Questions," Organization of Petroleum Exporting Countries, March 2009, www.opec.org/opec_web/static_files_project/media/downloads/publications/FAQ.pdf; "U.S. Imports by Country of Origin," US Energy Information Administration, http://tonto.eia.doe.gov/dnav/pet/pet_move_impcus_a2_nus_epc0_im0_mbblpd_a.htm (accessed July 29, 2009).

87. Trevor Morgan, *Reforming Energy Subsidies: Opportunities to Contribute to the Climate Change Agenda,* commissioned by the United Nations Environment Program (UNEP) and International Energy Agency (France, Switzerland, Japan: UNEP, 2002), 10. www.unep.org/pdf/PressReleases/Reforming_Energy_Subsidies2.pdf.

88. Dusty Horwitt, JD, "Free Pass for Oil and Gas: Environmental Protections Rolled Back As Western Drilling Surges: Oil and Gas Industry Exemptions," Environmental Working Group, March 2009, www.ewg.org/reports/Free-Pass-for-Oil-and-Gas/Oil-and-Gas-Industry-Exemptions.

89. *New York Times,* "Advertising Rates Effective January 1, 2009," *New York Times,* www.nytimes.whsites.net/mediakit/pdfs/newspaper/rates/2009/RateCard_Business09_EW6.pdf. The cost of an open 6.85 x 9.3 ad on the op-ed page of the *New York Times* (nationwide, weekday) was $53,455 in 2009.

90. Bernie Becker, "Baseball Team Clashes with Environmentalists over Oil Company Advertising," *New York Times,* July 27, 2008, www.nytimes.com/2008/07/27/us/27stadium.html; Peter Waldman, "Exxon vs. Obama," Portfolio.com, March 18, 2009, www.portfolio.com/business-news/portfolio/2009/03/18/Exxon-vs-the-Obama-Administration?page=4#page=4 (accessed July 14, 2009).

91. Waldman, "Exxon vs. Obama."

92. "Top Companies: Most Profitable," *Fortune Magazine,* May 4, 2009, http://money.cnn.com/magazines/fortune/fortune500/2009/full_list/; Jad Mouawad, "Oil

Giants Reluctant to Follow Obama's Green Lead," *New York Times*, April 8, 2009, www .nytimes.com/2009/04/08/business/energy-environment/08greenoil.html.

93. "USA: A Powerful Renewable Energy Policy Is Gaining Momentum," World Future Council, press release, July 31, 2009, www.worldfuturecouncil.org/feed_in _tariffs_in_us.html.

94. Ramez Naam, "Smaller, Cheaper, Faster: Does Moore's Law Apply to Solar Cells?" *Scientific American* Guest Blog, March 16, 2011, www.scientificamerican.com/ blog/post.cfm?id=smaller-cheaper-faster-does-moores-2011-03-15.

95. Miguel Mendonça, Stephen Lacey, and Frede Hvelplund, "Stability, Participation and Transparency in Renewable Energy Policy: Lessons from Denmark and the United States," *Policy and Society* 27, no. 4 (2009): 7. doi:10.1016/j.polsoc.2009.01.007.

96. Russ Christianson, "Danish Wind Co-ops Can Show Us the Way," Wind-Works.org, www.wind-works.org/articles/Russ%20Christianson%20NOW%20Article%20I.pdf.

97. "Ursula Sladek," Goldman Environmental Prize, www.goldmanprize.org/2011/ europe (accessed on April 22, 2011); Judy Dempsey and Jack Ewing, "Germany, in Re-versal, Will Close Nuclear Plants by 2022," *New York Times,* May 30, 2011, http://www .nytimes.com/2011/05/31/world/europe/31germany.html?scp=3&sq=germany%20 nuclear&st=cse.

98. *Renewable Energy Sources in Figures* (Berlin: Federal Ministry for the Environ-ment, 2010), 9. www.erneuerbare-energien.de/files/english/pdf/application/pdf/ broschuere_ee_zahlen_en_bf.pdf; *Energy Concept for an Environmentally Sound, Reliable and Affordable Energy Supply* (Berlin: Federal Ministry of Economics and Technology, 2010), 5. www.bmwi.de/English/Redaktion/Pdf/energy-concept,property=pdf,bereich =bmwi,sprache=en,rwb=true.pdf.

99. Renewables Insight, *PV Power Plants 2010* (Berlin: Solarpraxis, 2010), 10. www .pv-power-plants.com/fileadmin/user_upload/PVPP_2010_web.pdf; Herbert Girardet and Miguel Mendonça, *A Renewable World: A Report for the Future World Council* (Totnes, UK: Green Books, Ltd., 2009), 84; German Federal Environment Ministry, "Renewables' Contribution to Energy Supply in Germany Continued to Rise in 2010," Berlin, March 16, 2011, http://www.bmu.de/english/current_press_releases/pm/47124.php.

100. Renewable Energy Policy Network for the 21st Century (REN21), *Renewables 2010: Global Status Report* (Paris: REN21 Sectariat, 2010), 62, www.ren21.net/Portals/ 97/documents/GSR/REN21_GSR_2010_full_revised%20Sept2010.pdf.

101. Bill McKibben, "Global Warming's Terrifying New Math," *Rolling Stone,* July 2012.

102. David Cay Johnston, *Free Lunch: How the Wealthiest Americans Enrich Them-

*selves at Government Expense (and Stick You with the Bill)* (London: Portfolio Publishing, 2008), 275.

103. Robert Hopkins, *The Transition Handbook: From Oil Dependency to Local Resilience* (White River Junction, VT: First Chelsea Green, 2008), 186.

104. Hopkins, *The Transition Handbook*, 79.

105. Rowenna Davis, "Transition Towns—the Art of Resilience," *The New Internationalist* 430 (March 2010): 10.

106. "Initiatives Figures," Transition Network, www.transitionnetwork.org/initiatives (accessed April 6, 2011); Rob Hopkins, "Resilience Thinking," *Resurgence* 257 (November–December 2009): 15.

## Thought Trap 4

1. Philip Pullman, "The Shape of a Life and the Stories We Tell," *Do Good Lives Have to Cost the Earth?* eds. Andrew Simms and Joe Smith (London: Constable, 2008), 277.

2. E. O. Wilson, "Protect Biodiversity Hot Spots and the Rest Will Follow" (excerpts from lecture at Baldwin-Wallace College), *ScienceNews* 174, no. 13 (December 20, 2008), www.sciencenews.org/view/generic/id/39071/title/Comment__Protect_biodiversity _hot_spots_and_the_rest_will_follow.

3. Francesca Gino, Michael I. Norton, and Dan Ariely, "The Counterfeit Self: The Deceptive Costs of Faking It," *Psychological Science* 21, no. 5 (May 2010): 712–720.

4. John Horgan, "Quitting the Hominid Fight Club: The Evidence Is Flimsy for Innate Chimpanzee—Let Alone Human—Warfare," *Scientific American* Cross-Check (blog), June 29, 2010, www.scientificamerican.com/blog/post.cfm?id=quitting-the-hominid -fight-club-the-2010-06-29.

5. Horgan, "Quitting the Hominid Fight Club."

6. Charles Darwin, *The Descent of Man, and Selection in Relation to Sex* (London: Penguin, 2004), 143.

7. Yvonne L. Michael et al., "Health Behaviors, Social Networks and Healthy Aging: Cross-Sectional Evidence from the Nurses' Health Study," *Quality of Life Research* 8 (1999): 711–722.

8. Carol Schuck Scheiber, "Soul/Body: Loneliness Connects to Genes for Immune System," *Spirituality and Health* (March–April 2008), www.spiritualityhealth.com/spirit/content/loneliness-connects-genes-immune-system; see also Steve Cole et al., "Social Regulation of Gene Expression in Human Leukocytes," *Genome Biology* 8, no. 9 (2007): R189.

9. Adam Smith, *Theory of Moral Sentiments* (New York: Dover, 2006), II.ii.3.1.

10. For background, see Sarah Blaffer Hrdy, *Mothers and Others: The Evolutionary Origins of Mutual Understanding* (Cambridge, MA: Belknap Press, 2009); Michael Tomasello, *Why We Cooperate* (Cambridge, MA: MIT Press, 2009).

11. Michael Gurven, "To Give or Not to Give: The Behavioral Ecology of Human Food Transfers," *Behavioral and Brain Sciences* 27 (2004): 543–583.

12. Michael Alvard, "Good Hunters Keep Smaller Shares of Larger Pies," *Behavioral and Brain Sciences* 27 (2004): 560 (open peer commentary accompanying Gurven, "To Give or Not to Give").

13. Natalie Angier, "Why We're So Nice: We're Wired to Cooperate," *New York Times,* July 23, 2002, www.nytimes.com/2002/07/23/science/why-we-re-so-nice-we-re-wired -to-cooperate.html?src=pm.

14. Hrdy, *Mothers and Others,* 19–20, 27–28, 77–80, 179.

15. Richard Wilkinson and Kate Pickett, *The Spirit Level: Why More Equal Societies Almost Always Do Better* (London: Allen Lane-Penguin, 2009), 204–205; Marshall Sahlins, *The Use and Abuse of Biology* (Ann Arbor: University of Michigan Press, 1976), 100.

16. Michael Tomasello and Malinda Carpenter, "Shared Intentionality," *Developmental Science* 10, no. 1 (2007): 121–125.

17. Ágnes Melinda Kovács, Ernö Téglás, and Ansgar Denis Endress, "The Social Sense: Susceptibility to Others' Beliefs in Human Infants and Adults," *Science* 330, no. 6012 (December 2010): 1830–1834.

18. Marco Dondi, Francesca Simion, and Giovanna Caltran, "Can Newborns Discriminate Between Their Own Cry and the Cry of Another Newborn Infant?" *Developmental Psychology* 35, no. 2 (1999): 418–426.

19. Jules H. Masserman, Stanley Wechkin, and William Terris, "'Altruistic' Behavior in Rhesus Monkeys," *American Journal of Psychiatry* 121 (1964): 584–585. www.madison monkeys.com/masserman.pdf.

20. John Drury and Stephen D. Reicher, "Crowd Control," *Scientific American Mind,* November 2010, 60.

21. Elsa Youngsteadt, "The Secret to Happiness? Giving," *ScienceNOW,* March 20, 2008, 2. http://news.sciencemag.org/sciencenow/2008/03/20-02.html.

22. Tim Jackson, *Prosperity Without Growth: Economics for a Finite Planet* (London: Earthscan, 2009), 37.

23. Adam Phillips and Barbara Taylor, *On Kindness* (New York: Farrar, Straus and Giroux, 2009), 5, 51.

24. Allan Luks and Peggy Payne, *The Healing Power of Doing Good* (New York: Ballantine, 1992), 81.

25. Adam Smith, *Theory of Moral Sentiments*, II.ii.1.5.

26. Sarah F. Brosnan and Frans B. M. de Waal, "Monkeys Reject Unequal Pay," *Nature* 425 (September 18, 2003): 297–299.

27. Antonio Damasio, *Looking for Spinoza: Joy, Sorrow, and the Feeling Brain* (Orlando, FL: Harcourt, 2003), 160.

28. Robert Kurzban, Peter DeScioli, and Erin O'Brien, "Audience Effects on Moralistic Punishment," *Evolution and Human Behavior* 28, no. 2 (2007): 75–84.

29. Melissa Bateson, Daniel Nettle, and Gilbert Roberts, "Cues of Being Watched Enhance Cooperation in a Real-World Setting," *Biological Letters* 2, no. 3 (2006): 412–414. Available at http://rsbl.royalsocietypublishing.org/content/2/3/412.

30. Martin A. Nowak, Karen M. Page, and Karl Sigmund, "Fairness Versus Reason in the Ultimatum Game," *Science* 289, no. 5485 (2000): 1773.

31. Elizabeth Tricomi et al., "Neural Evidence for Inequality-Averse Social Preferences," *Nature* 463 (February 2010): 1089–1091; Melinda Moyer, "You're Happy, I'm Happy: Biology Plays a Role in Our Aversion to Inequity," *Scientific American Mind,* 13.

32. Richard Wilkinson, *The Impact of Inequality: How to Make Sick Societies Healthier* (New York: New Press, 2005), 57, 69, 75.

33. Erich Fromm, *The Heart of Man* (New York: Harper & Row, 1964), 26.

34. Erich Fromm, *The Anatomy of Human Destructiveness* (New York: Holt, Rinehart & Winston, 1973), 264.

35. Alison Gopnik, *The Philosophical Baby: What Children's Minds Tell Us About Truth, Love, and the Meaning of Life* (New York: Farrar, Strauss and Giroux, 2009), 87.

36. Adam Gorlick, "For Kids, Altruism Comes Naturally, Psychologist Says," *Stanford Report*, November 5, 2008, http://news.stanford.edu/news/2008/november5/tanner-110508.html?view=print.

37. Felix Warneken and Michael Tomasello, "Altruistic Helping in Human Infants and Young Chimpanzees," *Science* 311, no. 5765 (March 2006): 1301–1303.

38. Gorlick, "For Kids, Altruism Comes Naturally, Psychologist Says."

39. Ellen Langer, *Counterclockwise: Mindful Health and the Power of Possibility* (New York: Ballantine Books, 2009), 3–5.

40. "Our Culture: Fostering Associate Engagement with the Gallup Q12 Survey," Interface Global, www.interfaceglobal.com/Company/Culture/Associate-Engagement.aspx (accessed April 5, 2011).

41. Maarten Vansteenkiste et al., "Motivating Learning, Performance, and Persistence: The Synergistic Effects of Intrinsic Goal Contents and Autonomy-Supportive Contexts," *Journal of Personality and Social Psychology* 87, no. 2 (August 2004): 246–260.

42. Gopnik, *The Philosophical Baby*, 6–9.

43. Joe Palca, "Tiny Water Flea Clocks in Record Number of Genes," *Morning Edition*, National Public Radio, February 4, 2011, www.npr.org/2011/02/04/133466183/tiny-water-flea-clocks-in-record-number-of-genes.

44. Stanley Milgram, *Obedience to Authority* (New York: Harper & Row, 1974).

45. Emily Anthes, "Their Pain, Our Gain," *Scientific American Mind*, November 2010, 39–41. See also Russell Spears and Colin Wayne Leach, "Intergroup Schadenfreude: Conditions and Consequences," in *The Social Life of Emotions*, ed. Larissa Z. Tiedens and Colin Wayne Leach (Cambridge, UK: Cambridge University Press, 2004).

46. World Bank, *World Development Indicators 2010* (Washington, DC: International Bank for Reconstruction and Development/World Bank, 2010), 94–96. http://data.worldbank.org/sites/default/files/wdi-final.pdf; "Income Share Held by Lowest 20%," World Development Indicators, http://data.worldbank.org/indicator/SI.DST.FRST.20 (accessed February 28, 2011).

47. Richard Wilkinson and Kate Pickett, *The Spirit Level: Why More Equal Societies Almost Always Do Better* (London: Allan Lane, 2009), 67.

48. "Mental Health: Depression," World Health Organization, www.who.int/mental_health/management/depression/definition/en/; "Global Burden of Disease: 2004 Update (2008)," World Health Organization, www.who.int/healthinfo/global_burden_disease/GBD_report_2004update_part4.pdf, 43.

49. Wilkinson and Pickett, *The Spirit Level*.

50. Vivian Gussin Paley, *You Can't Say You Can't Play* (Cambridge, MA: Harvard University Press, 1993). See also interview with Vivian Gussin Paley by Ira Glass, *This American Life*, Chicago Public Radio, August 22, 2009, www.thisamericanlife.org/radio-archives/episode/27/the-cruelty-of-children.

51. Susan Engle and Marlene Sandstrom, "There's Only One Way to Stop a Bully," *New York Times*, July 22, 2010, www.nytimes.com/2010/07/23/opinion/23engel.html.

52. Victoria E. Sturtevant and Jonathan I. Lange, *Applegate Partnership Case Study: Group Dynamics and Community Context* (Ashland, OR: Southern Oregon State College, 1996), 1, 7–11. http://soda.sou.edu/awdata/030516a1.pdf.

53. Lydialyle Gibson, "Mirrored Emotion," *University of Chicago Magazine* 98, no. 4 (April 2006), http://magazine.uchicago.edu/0604/features/emotion.shtml.

54. See the website of the Education for Sustainable Living Program, University of California, Santa Cruz, at http://eslp.enviroslug.org.

55. Tom Crompton and Tim Kasser, *Meeting Environmental Challenges: The Role of Human Identity* (London: WWF-UK, 2009), 18–22.

56. Crompton and Kasser, *Meeting Environmental Challenges*, 14, 17–22.

57. National Council for Science and the Environment, David E. Blockstein, and Leo Wiegman, *The Climate Solutions Consensus: What We Know and What to Do About It* (Washington, DC: Island Press, 2010), 174.

58. Robert B. Cialdini, "Don't Throw in the Towel: Use Social Influence Research," *Observer* 18, no. 4 (April 2005), www.psychologicalscience.org/observer/getArticle .cfm?id=1762.

59. "Killed by the Messenger," interview with Robert Cialdini by Steve Curwood, *Living on Earth*, National Public Radio, December 17, 2010. www.loe.org/shows/shows .htm?programID=10-P13–00051.

60. National Council for Science and the Environment, Blockstein, and Wiegman, *Climate Solutions Consensus*, 174.

61. Gopnik, *The Philosophical Baby*, 82–85.

## Thought Trap 5

1. Paul Reynolds, "'Freedom' at the Heart of Bush Foreign Policy," BBC News, February 7, 2005, http://news.bbc.co.uk/2/hi/americas/4232241.stm.

2. Amrisha Vaish, Malinda Carpenter, and Michael Tomasello, "Young Children Selectively Avoid Helping People with Harmful Intentions," *Child Development* 81, no. 6 (November–December 2010): 1661–1669.

3. Intergovernmental Panel on Climate Change, *Climate Change 2007: Impacts, Adaptation and Vulnerability* (Cambridge, UK: Cambridge University Press, 2007), www.ipcc.ch/publications_and_data/ar4/wg2/en/contents.html.

4. "Where Carbon Is Taxed," Carbon Tax Center, September 1, 2010, www.carbontax .org/progress/where-carbon-is-taxed; Lester Brown, *Plan B 4.0: Mobilizing to Save Civilization* (New York: W. W. Norton, 2009), 244–247.

5. Brown, *Plan B 2.0*, 229.

6. Angelika Pullen, Liming Qiao, and Steve Sawyer, eds., *Global Wind 2009 Report* (Brussels: Global Wind Energy Council, 2010), 9–10. www.gwec.net/fileadmin/ documents/Publications/Global_Wind_2007_report/GWEC_Global_Wind_2009_Report _LOWRES_15th.%20Apr..pdf.

7. US Energy Information Administration, *Emissions of Greenhouse Gases in the United States 2008* (Washington, DC: US Department of Energy, 2009), 16, 24. Available at www.eia.doe.gov/oiaf/1605/ggrpt/pdf/0573%282008%29.pdf; David Austin, "Climate-Change Policy and $CO_2$ Emissions from Passenger Vehicles" (Economic and Budget Issue Brief, Congressional Budget Office, October 6, 2008), www.cbo.gov/ftpdocs/98xx/doc9830/10–06-ClimateChange_Brief.pdf.

8. "Where Carbon Is Taxed," Carbon Tax Center.

9. Dr. David Pimentel, "Soil Erosion: A Food and Environmental Threat," *Environment Development and Sustainability* 8, no. 1 (2006): 131, citing the following USDA reports: Natural Resources Conservation Service, "Changes in Average Annual Soil Erosion by Water on Cropland and CRP Land, 1992–1997," USDA, www.nrcs.usda.gov/technical/NRI/maps/meta/m5061.html (last modified December 7, 2000); Natural Resources Conservation Service, "Changes in Average Annual Soil Erosion by Wind on Cropland and CRP Land, 1992–1997," USDA, www.nrcs.usda.gov/technical/NRI/maps/meta/m5066.html (last modified December 7, 2000).

10. For a summary of the work of leaders in the field of soil biology, see Arlene Tugel, Ann Lewandowski, and Deb Happe-vonArb, eds., *Soil Biology Primer*, rev. ed. (Ankeny, IA: Soil and Water Conservation Society, 2000), http://soils.usda.gov/sqi/concepts/soil_biology/biology.html.

11. "Farming Must Change to Feed the World, FAO Expert Urges More Sustainable Approach," Food and Agriculture Organization of the United Nations, news release, February 4, 2009, www.fao.org/news/story/en/item/9962/icode.

12. Sarah Anderson and John Cavanagh, "Top 200: The Rise of Corporate Global Power," Institute for Policy Studies, December 4, 2000, www.ips-dc.org/reports/top_200_the_rise_of_corporate_global_power.

13. Adam Winkler, "Corporate Personhood and the Rights of Corporate Speech," *Seattle University Law Review* 30 (2007): 863–864.

14. This is according to the official court syllabus in the *United States Reports, Santa Clara County v. Southern Pacific Railroad Company* 118 U.S. 394, 396 (1886).

15. Thom Hartmann, *Unequal Protection: The Rise of Corporate Dominance and the Theft of Human Rights* (Emmaus, PA: Rodale, 2002), 112.

16. Winkler, "Corporate Personhood and the Rights of Corporate Speech," 865.

17. *Federal Communication Commission v. AT&T Corp.*, 562 U.S. ___ (2011), Supreme Court of the United States, www.supremecourt.gov/opinions/10pdf/09–1279.pdf.

18. Adam Liptak, "Justices, 5–4, Reject Corporate Spending Limit," *New York Times*,

January 21, 2010, www.nytimes.com/2010/01/22/us/politics/22scotus.html?ref=campaignfinance.

19. Dan Eggen, "Poll: Large Majority Opposes Supreme Court's Decision on Campaign Financing," *Washington Post*, February 17, 2010, www.washingtonpost.com/wp-dyn/content/article/2010/02/17/AR2010021701151.html.

20. *Dartmouth College v. Woodward*, 17 U.S. 518 at 636 (1819).

21. Simon Johnson, "The Quiet Coup," *The Atlantic* (May 2009), www.theatlantic.com/magazine/archive/2009/05/the-quiet-coup/7364.

22. "Employment, Hours and Earnings from the Current Employment Statistics Survey," Bureau of Labor Statistics, www.bls.gov/ces (accessed March 4, 2011).

23. Paul Krugman, "School for Scoundrels," review of *Myth of the Rational Market*, by Justin Fox, *New York Times Book Review*, August 6, 2009, www.nytimes.com/2009/08/09/books/review/Krugman-t.html.

24. Interview with Alex Blumberg and Adam Davidson by Ira Glass, "The Giant Pool of Money," *This American Life*, Program No. 355, May 9, 2008, www.thisamericanlife.org/sites/default/files/355_transcript.pdf.

25. Hearings on: Wall Street Fraud and Fiduciary Duties: Can Jail Time Serve As an Adequate Deterrent for Willful Violations? Before the Subcommittee on Crime, Senate Judiciary Committee, 111th Congress (2010) (statements by James K. Galbraith and Lloyd M. Bentsen, Junior Chair in Government/Business Relations, Lyndon B. Johnson School of Public Affairs, University of Texas at Austin).

26. Frank Ahrens, "Post Exclusive: Greenspan Says Bad Big Banks Should Be Busted Up," *Washington Post* Economy Watch (blog), October 30, 2009, http://voices.washingtonpost.com/economy-watch/2009/10/exclusive_greenspan_says_geith.html.

27. Karl Polanyi, *The Great Transformation* (Boston: Beacon Press, 1944).

28. Francis X. Sutton et al., *The American Business Creed* (Cambridge, UK: Cambridge University Press, 1956), 64–65, as quoted in Sanford M. Jacoby, "Finance and Labor: Perspectives on Risk, Inequality and Democracy" (working paper, University of California, Los Angeles, 2008), 25. http://papers.ssrn.com/sol3/papers.cfm?abstract_id=1020843.

29. Clyde Prestowitz, *The Betrayal of American Prosperity* (New York: Free Press, 2010), 193. The author served in the Reagan administration.

30. Edward S. Mason, "Corporation," in *International Encyclopedia of the Social Sciences*, ed. David L. Sills (New York: Macmillan, 1968), 397.

31. Hardik Savalia (B Lab Core Team member, B Corporation), in discussion with the author, September 17, 2009.

32. Michael E. Porter and Mark R. Kramer, "The Big Idea: Creating Shared Value," *Harvard Business Review* (January–February 2011): 4.

33. Alexandra Lejoux, National Association of Corporate Directors, personal communication with author, 2004.

34. Pennsylvania Consolidated Statutes, Title 15, Corporations and Unincorporated Associations, Pa.C.S. 1715 (2004). The statute says, "In discharging the duties of their respective positions, the board of directors, committees of the board and individual directors of a business corporation may, in considering the best interests of the corporation, consider to the extent they deem appropriate: The effects of any action upon any or all groups affected by such action, including shareholders, employees, suppliers, customers and creditors of the corporation, and upon communities in which offices or other establishments of the corporation are located. . . . The board of directors, committees of the board and individual directors shall not be required, in considering the best interests of the corporation or the effects of any action, to regard any corporate interest or the interests of any particular group affected by such action as a dominant or controlling interest or factor."

35. "Public Policy," B Corporation, www.bcorporation.net/index.cfm/nodeID/BE2FD378-D039-4D35-90A7-48B824BCAC78/fuseaction/content.page (accessed April 19, 2011).

36. "Public Policy," B Corporation.

37. Tina Rosenberg, "A Scorecard for Companies with a Conscience," *New York Times* Opinionator Blog, April 11, 2011, http://opinionator.blogs.nytimes.com/2011/04/11/a-scorecard-for-companies-with-a-conscience/?partner=rss&emc=rss.

38. This estimate is taken from latest data available: 1998 to 2004. Katherine Kobe, *The Small Business Share of GDP, 1998–2004* (Small Business Research Summary 299, Small Business Administration, Office of Advocacy, April 2007), 1. www.smallbusinessnotes.com/pdf/rs299tot.pdf.

39. Marc Orlitzky, Frank L. Schmidt, and Sara L. Rynes, "Corporate Social and Financial Performance: A Meta-Analysis," *Organization Studies* 24, no. 3 (2003): 403–441; Andrew White and Matthew Kiernan, *Corporate Environmental Governance* (Bristol, UK: Environment Agency, 2004), http://publications.environment-agency.gov.uk/pdf/GEHO0904BKFE-e-e.pdf?lang=_e.

40. Daniel Mahler et al., "'Green' Winners: The Performance of Sustainability-Focused Companies During the Financial Crisis," A. T. Kearney, 2009, www.atkearney.com/images/global/pdf/Green_winners.pdf.

41. Doug Pibel, "Communities Take Power," *Yes! Magazine* 43 (fall 2007): 25–28. www.yesmagazine.org/issues/stand-up-to-corporate-power/communities-take-power.

42. "Statement from Shelly Gobeille, Protect Our Water and Wildlife Resources, at the Nestle Waters North America Headquarters," Corporate Accountability International, www.stopcorporateabuse.org/statement-shelly-gobeille-protect-our-water-and -wildlife-resources-nestle-waters-north-america-headq+Shelly+Gobeille+outrage &cd=1&hl=en&ct=clnk&gl=us&client=firefox-a.

43. Daniel Moss and Brooke Jarvis, "Signs of Life: Defending the Right to Water— Maine Towns Fight Back," *Yes! Magazine* 50 (summer 2009): 6–7. www.yesmagazine .org/issues/the-new-economy/signs-of-life-defending-the-right-to-water.

44. Ian Urbina, "E.P.A. Struggles to Police Drilling for Gas," *New York Times*, March 3, 2011, www.nytimes.com/2011/03/04/us/04gas.html?pagewanted=1&_r=1.

45. Mari Margil and Ben Price, "Pittsburgh Bans Natural Gas Drilling," YesMagazine.org, November 16, 2010, www.yesmagazine.org/people-power/pittsburg -bans-natural-gas-drilling.

46. Miguel Mendonça, David Jacobs, and Benjamin Sovacool, *Powering the Green Economy: The Feed-in Tariff Handbook* (London: Earthscan, 2009); Miguel Mendonça, personal communication with author, December 10, 2009.

47. Interview with Ed Regan by Steve Curwood, *Living on Earth*, National Public Radio, May 22, 2009, www.loe.org/shows/segments.htm?programID=09-P13–00021 &segmentID=4. See also "Leading the Nation: GRU's Solar Feed-in-Tariff," Gainesville Regional Utilities, news release, February 6, 2009, https://www.gru.com/AboutGRU/ NewsReleases/Archives/Articles/news-2009-02–06.jsp.

48. Bernadette Del Chiaro, *California's Solar Cities: Leading the Way to a Clean Energy Future* (Los Angeles: Environment California, 2009), 8. www.environmentcalifornia .org/uploads/YM/3W/YM3W81JComzW53sx1fgAiw/Californias-Solar-Cities.pdf; Felicity Barringer, "With Push Toward Renewable Energy, California Sets Pace for Solar Power," *New York Times*, July 17, 2009, www.nytimes.com/2009/07/16/science/ earth/16solar.html.

49. United States Conference of Mayors, "List of Participating Mayors," US Conference of Mayors Climate Protection Center, www.usmayors.org/climateprotection/list.asp.

50. David Gorn, "San Francisco Plastic Bag Ban Interests Other Cities," National Public Radio, www.npr.org/templates/story/story.php?storyId=89135360 (accessed March 27, 2008).

51. Union of Concerned Scientists (UCS), "National Clean Car Standards Factsheet— National Clean Vehicle Program: Model Year 2012–2016 Standards," UCS, April 2010, www.ucsusa.org/assets/documents/clean_vehicles/National-Clean-Car-Standards -Fact-Sheet.pdf.

52. World Future Council, *Policies to Change the World: Energy Sufficiency—Eight Policies Towards the Sustainable Use of Energy* (Hamburg: World Future Council, 2009), 7. See also "Philippines to Ban Incandescent Bulbs," Fox News, February 5, 2008, www.foxnews.com/wires/2008Feb05/0,4670,PhilippinesBulbBan,00.html.

53. Steven Mufson, "In Energy Conservation, California Sees Light," *Washington Post*, February 17, 2007, www.washingtonpost.com/wp-dyn/content/article/2007/02/16/AR2007021602274_pf.html.

54. Ernst von Weizsacker, Amory B. Lovins, and L. Hunter Lovins, *Factor Four: Doubling Wealth, Halving Resource Use* (London: Earthscan, 1997), 159–160.

55. "Decoupling Policies," Pew Center on Global Climate Change, www.pewclimate.org/what_s_being_done/in_the_states/decoupling (last modified February 10, 2011).

56. Adrienne Kandel, Margaret Sheridan, and Patrick McAuliffe, "A Comparison of per Capita Electricity Consumption in the United States and California" (paper presented at the 2008 Summer Study on Energy Efficiency in Buildings, Panel 8, Paper 12, Pacific Grove, California, 2008), 124–125. www.aceee.org/sites/default/files/publications/proceedings/SS08_Panel8_Paper12.pdf; Climate Analysis Indicators Tool (CAIT) Version 7.0. (Total GHG Emissions in 2005; accessed January 16, 2011), http://cait.wri.org/; Air Resources Board, "Trends in California Greenhouse Gas Emissions for 2000 to 2008: By Category as Defined in the Scoping Plan," California Environmental Protection Agency, May 28, 2010, 1, http://www.arb.ca.gov/cc/inventory/data/tables/ghg_inventory_trends_00–08_2010-05-12.pdf.

57. "Key Events in the History of Air Quality in California," California Environmental Protection Agency Air Resources Board, www.arb.ca.gov/html/brochure/history.htm (last modified January 13, 2011).

58. "Profile of the Charles River," Charles River Watershed Association, www.crwa.org/cr_history.html (accessed January 4, 2011). See also "Clean Charles River Initiative," Environmental Protection Agency New England, www.epa.gov/NE/charles/initiative.html (accessed March 13, 2011).

59. Charles Duhigg, "Clean Water Laws Are Neglected, at a Cost in Suffering," *New York Times*, September 12, 2009, www.nytimes.com/2009/09/13/us/13water.html.

60. Elinor Ostrom, *Governing the Commons: The Evolution of Institutions for Collective Action* (Cambridge, UK: Cambridge University Press, 1990), 58–65.

61. "The United Nations Convention on the Laws of the Sea: A Historical Perspective," Division for Ocean Affairs and the Law of the Sea, www.un.org/Depts/los/convention_agreements/convention_historical_perspective.htm#Settlement%20of%20Disputes.

62. "Chronological Lists of Ratifications of Accessions and Successions to the

Convention and the Related Agreements," Division for Ocean Affairs and the Law of the Sea, www.un.org/Depts/los/reference_files/chronological_lists_of_ratifications.htm# The%20United%20Nations%20Convention%20on%20the%20Law%20of%20the%20Sea (accessed April 29, 2011).

63. Sylvia A. Earle, *The World Is Blue: How Our Fate and the Ocean's Are One* (Washington, DC: National Geographic, 2009), 210.

64. Ransom A. Myers and Boris Worm, "Rapid Worldwide Depletion of Predatory Fish Communities," *Nature* 423 (May 2003): 280–283. www.nature.com/nature/journal/ v423/n6937/abs/nature01610.html; Amy Matthews-Amos and Ewwan A. Berntson, *Turning up the Heat: How Global Warming Threatens Life in the Sea* (Washington, DC: WWF, 1999), www.worldwildlife.org/who/media/press/1999/WWFBinaryitem13089 .pdf; John Roach, "Source of Half Earth's Oxygen Gets Little Credit," National Geographic News, June 7, 2004, http://news.nationalgeographic.com/news/2004/06/ 0607_040607_phytoplankton.html.

65. Clifford Krauss et al., "As Polar Ice Turns to Water, Dreams of Treasure Abound," *New York Times*, October 10, 2005. http://www,nytimes.com/2005/10/10/science/ 10artic.html

66. Mark Dowie, "Conservation Refugees," *Orion* (November–December 2005), www.orionmagazine.org/index.php/articles/article/161. See also "Global Facts About Marine Protected Areas and Marine Reserves," Protect Planet Ocean, www .protectplanetocean.org/collections/introduction/introbox/globalmpas/story.html (accessed February 24, 2011).

67. Steve Connor, "Climate Change Is Killing the Oceans' Microscopic 'Lungs,'" *Independent*, December 7, 2006, www.independent.co.uk/environment/climate-change/ climate-change-is-killing-the-oceans-microscopic-lungs-427402.html.

68. Kristin Kolb, "Canada Extends Conservation Area to Sea Floor," *Yes! Magazine* 55 (fall 2010), www.yesmagazine.org/issues/a-resilient-community/canada-extends -conservation-area-to-seafloor.

69. "Ecuador Yasuni ITT Trust Fund: Terms of Agreement," UN Development Program, Multi-Donor Trust Fund Office Gateway, July 28, 2010, http://mdtf.undp.org/ yasuni; Lorna Howarth, "Leaving Oil in the Ground," *Resurgence* 263 (November– December 2010): 6.

70. Matt Finer and Pamela Martin, "The Current State of the Yasuni-ITT Initiative (Part III)," *Globalist*, June 25, 2010, www.theglobalist.com/StoryId.aspx?StoryId=8535; "Ecuador Yasuni ITT Trust Fund: Terms of Agreement," United Nations Development Program, Multi-Donor Trust Fund Office Gateway, July 28, 2010, http://mdtf.undp.org/yasuni.

71. Hylton Murray-Philipson, "From Copenhagen to Cancun," *Resurgence* 263 (November–December 2010): 12, www.resurgence.org/magazine/article3223-from -copenhagen-to-cancun.html. His calculation is based on the $32 trillion loss reported between October 2007 and March 2009.

72. Bina Agarwal, *Gender and Green Governance: The Political Economy of Women's Presence Within and Beyond Community Forestry* (London: Oxford University Press, 2010), 11, 369; Bina Agarwal, personal communication with author, May 2011.

73. Food and Agriculture Organization of the United Nations (FAO), *Global Forest Resources Assessment 2010* (Rome: FAO, 2010), 21. www.fao.org/docrep/013/i1757e/ i1757e.pdf.

74. Thomas L. Friedman, "(No) Drill, Baby, Drill," *New York Times*, April 12, 2009, www.nytimes.com/2009/04/12/opinion/12friedman.html.

75. "Ecuador Adopts New Constitution—with CELDF Rights of Nature Language," Community Environmental Legal Defense Fund, news release, September 28, 2008, www.celdf.org/rights-of-nature.

76. Jay Walljasper, "Vision: Kenya Enshrines the Environment in Its Constitution— This Should Be Our Future," Alternet, January 3, 2011, www.alternet.org/environment/ 149401/vision%3A_kenya_enshrines_the_environment_in_its_constitution_–_this_ should_be_our_future; Jeffrey Gettleman, "Kenyans Approve New Constitution," *New York Times*, August 5, 2010, http://www.nytimes.com/2010/08/06/world/africa/ 06kenya.html.

77. Peter Burdon, "The Rights of Nature: Reconsidered," *Australian Humanities Review* 49 (2010), http://epress.anu.edu.au/apps/bookworm/view/Australian+Humanities +Review+-+Issue+49,+2010/1851/05.xhtml; Ben Price, (projects director, Community Environmental Legal Defense Fund), e-mail communication with the author, December 16, 2010.

78. Maja Göpel, *Guarding Our Future: How to Include Future Generations in Policy Making* (Hamburg: World Future Council, 2010), 5. www.worldfuturecouncil.org/fileadmin/ user_upload/PDF/brochure_guardian3.pdf.

79. For more about "future justice," visit the World Future Council at www .worldfuturecouncil.org/future_justice.html.

## Thought Trap 6

1. "World Urbanization Prospects: The 2007 Revision Population Database," United Nations Department of Economic and Social Affairs, http://esa.un.org/unup. This database defines "urban population" as the total population living in areas termed

"urban" by a given country. Typically, the population living in towns of 2,000 or more or in national and provincial capitals is classified as "urban."

2. Steve Connor, "Children Better at Recognising Pokémon Characters Than British Wildlife," *Independent*, March 29, 2002, www.independent.co.uk/news/uk/home-news/children-better-at-recognising-pokemon-characters-than-british-wildlife-655781.html.

3. Richard Louv, *Last Child in the Woods: Saving Our Children from Nature-Deficit Disorder* (New York: Algonquin Books, 2005), 10.

4. E. O. Wilson, "Biophilia and the Conservation Ethic," in *The Biophilia Hypothesis*, ed. Stephen Kellert and Edward O. Wilson (Washington, DC: Island, 1993), 31.

5. "Hard Times Lawn & Garden Survey Says More People Saving with Do-It-Yourself Lawn & Garden Care," National Gardening Association, news release, August 6, 2010, http://assoc.garden.org/press/press.php?q=show&id=3324&pr=pr_research.

6. American Community Gardening Association, "Frequently Asked Questions," www.communitygarden.org/learn/faq.php (accessed February 28, 2011).

7. "Urban Agriculture for Sustainable Poverty Alleviation and Food Security," Food and Agriculture Organization of the United Nations, 2008, www.rlc.fao.org/es/agricultura/aup/pdf/upa.pdf, 22.

8. Emily Friedman, "Meanwhile Back at the Ranch," *Health Forum Journal* (November–December 2000), 6, quoted in Katherine H. Brown, "Urban Agriculture and Community Food Security in the United States: Farming from the City Center to the Urban Fringe" (Venice, CA: Community Food Security Coalition, 2002), 5. www.foodsecurity.org/urbanagpaper.pdf.

9. Andrea Faber Taylor et al., "Growing Up in the Inner City: Green Spaces As Places to Grow," *Environment and Behaviour* 30, no. 1 (1998): 3–27; Andrea Faber Taylor, Frances E. Kuo, and William C. Sullivan, "Coping with ADD: The Surprising Connection to Green Play Settings," *Environment and Behaviour* 33 (2001): 54–77.

10. Carolyn M. Tennessen and Bernadine Cimprich, "Views to Nature: Effects on Attention," *Journal of Environmental Psychology* 15, no. 1 (1995): 77–85.

11. Jules Pretty et al., "The Mental and Physical Health Outcomes of Green Exercise," *International Journal of Environmental Health Research* 15, no. 5 (October 2005): 319–337. www.essex.ac.uk/bs/bs_staff/pretty/IJEHR%20Green%20exercise%20%28Pretty%20et%20al%202005%29.pdf.

12. Roger S. Ulrich, "View Through a Window May Influence Recovery from Surgery," *Science* 224, no. 4647 (1984): 420–421. www.sciencemag.org/content/224/4647/420.abstract.

13. Gregory B. Diette et al., "Distraction Therapy with Nature Sights and Sounds Reduces Pain During Flexible Bronchoscopy: A Complementary Approach to Routine Analgesia," *Chest* 123, no. 3 (2003): 941–948. http://chestjournal.chestpubs.org/content/123/3/941.full.pdf+html.

14. Lorna Howarth, "Care Farming," *Resurgence Magazine* 258 (January–February 2010): 7.

15. S. Richard Mitchell and Frank Popham, "Effect of Exposure to Natural Environment on Health Inequalities: An Observational Population Study," *The Lancet* 372, no. 9650 (November 2008): 1655–1660.

16. Richard Louv, "The Powerful Link Between Conserving Land and Preserving Human Health," Field Notes from the Future, July 1, 2007, www.childrenandnature.org/blog/?p=36.

17. Sarah Goodwin, "Frumkin's $R_x$: Intense Exposure to Natural Elements," *Emory Report*, April 9, 2001, www.emory.edu/EMORY_REPORT/erarchive/2001/April/erApril.9/4_9_01frumkin.html.

18. Jane Brody, "Head Out for a Daily Dose of Green Space," *New York Times*, November 30, 2010, www.nytimes.com/2010/11/30/health/30brody.html.

19. "Chicago City Hall," Green Roofs Projects Database, www.greenroofs.com/projects/pview.php?id=21 (accessed February 7, 2011).

20. Marian Burros, "Urban Farming, a Bit Closer to the Sun," *New York Times*, June 16, 2009, www.nytimes.com/2009/06/17/dining/17roof.html.

21. Natalie Hope McDonald, "Sweet Success," Grid, September 27, 2010, www.gridphilly.com/grid-magazine/2010/9/27/sweet-success.html.

22. Steven McFadden, "Community Farms in the 21st Century—Poised for Another Wave of Growth?" Rodale Institute, http://newfarm.rodaleinstitute.org/features/0104/csa-history/part1.shtml (accessed December 14, 2010). See also "Community Supported Agriculture," Local Harvest, www.localharvest.org/csa.

23. "Farmers Market Growth, 1994–2010," US Department of Agriculture, Agricultural Marketing Service, www.ams.usda.gov/AMSv1.0/ams.fetchTemplateData.do?template=TemplateS&navID=WholesaleandFarmersMarkets&leftNav=WholesaleandFarmersMarkets&page=WFMFarmersMarketGrowth&description=Farmers%20Market%20Growth&acct=frmrdirmkt (last modified August 4, 2010).

24. Brian Halweil, *Home Grown: The Case for Local Food in a Global Market*, Worldwatch Paper 163, ed. Thomas Prugh (Washington, DC: Worldwatch Institute, November 2002): 13. Available at www.worldwatch.org/system/files/EWP163.pdf.

25. For more information, visit Local Harvest at www.localharvest.org.

26. Karla Adam, "English Town Digs Up Lots of Space to Grow," *Washington Post*, August 16, 2009, www.washingtonpost.com/wp-dyn/content/article/2009/08/15/AR2009081502031.html.

27. Nathan McClintock and Jenny Cooper, *Cultivating the Commons: An Assessment of the Potential for Urban Agriculture on Oakland's Public Land* (Oakland, CA: Institute for Food and Development Policy, 2010), 13, 28. www.urbanfood.org/docs/Cultivating_the_Commons2010.pdf. For more information on the Oakland Food Policy Council, visit www.oaklandfood.org/home.

28. Kären Haley, director, Indianapolis Office of Sustainability, e-mail communication with author, December 3, 2010.

29. Kathryn Colasanti and Michael Hamm, "Assessing the Local Food Supply Capacity of Detroit, Michigan," *Journal of Agriculture, Food Systems, and Community Development* 1, no. 2 (2010): 51–58. www.agdevjournal.com/attachments/137_JAFSCD_Assessing_Food_Supply_Capacity_Detroit_Nov-2010.pdf.

30. Chicago Jobs Council, "Sprouting Roots: Social Enterprises in Landscaping & Horticulture," *Industry Insider* (spring 2007): 11–13. www.cjc.net/industry_insider/industry_insider_sp07_sprouting.php.

31. Growing Home, *Growing Home Annual Report 2008* (Chicago: Growing Home, 2008), www.growinghomeinc.org/storage/AnnualReport2008.pdf.

32. Shannon Hayes, "Will Allen's Growing Power: Growing Much More Than Food," *Agriview*, February 4, 2010, www.agriview.com/articles/2010/02/04/features/feature01.txt.

33. Roger Bybee, "Growing Power in an Urban Food Desert," *Yes! Magazine*, February 13, 2009, www.yesmagazine.org/issues/food-for-everyone/growing-power-in-an-urban-food-desert.

34. Abdul Alim Muhammad, "Farrakhan Visits Growing Power: A Meeting of Giants," Final Call, December 12, 2010, www.finalcall.com/artman/publish/National_News_2/article_7492.shtml; Shannon Hayes, "Will Allen's Growing Power: Growing Much More Than Food," AgriView, February 4, 2010, www.agriview.com/articles/2010/02/04/features/feature01.txt.

35. Barbara Miner, "An Urban Farmer Is Rewarded for His Dream," *New York Times*, October 1, 2008, www.nytimes.com/2008/10/01/dining/01genius.htm.

36. Interview with Will Allen by Jacki Lyden, "Urban Farming," *On Point*, National Public Radio, July 30, 2009. www.onpointradio.org/2009/07/urban-farming.

37. Antoine de Saint-Exupéry, *Wisdom of the Sands: Spiritual Science* (Chicago: University of Chicago Press, 1979).

38. "School Garden Program Overview," California Department of Education, www.cde.ca.gov/Ls/nu/he/gardenoverview.asp (accessed March 23, 2010).

39. American Institutes for Research, *Effects of Outdoor Education Programs for Children in California* (Washington, DC: American Institutes for Research, 2005), 29. www.air.org/files/Outdoorschoolreport.pdf.

40. Hillary L. Burdette and Robert C. Whitaker, "Resurrecting Free Play in Young Children: Looking Beyond Fitness and Fatness to Attention, Affiliation, and Affect," *Archives of Pediatrics and Asolescent Medicine* 159 (2005): 46–50.

41. Jonathan Rosen, "The Natural Man," review of *The Wilderness Warrior: Theodore Roosevelt and the Crusade for America* by Douglas Brinkley, *New York Times Book Review*, August 6, 2009, www.nytimes.com/2009/08/09/books/review/Rosen-t.html.

42. For an overview, see the Forest Education Initiative at www.foresteducation.org.

## Thought Trap 7

1. James Lovelock, "Lovelock: 'We Can't Save the Planet,'" interview by John Humphreys, BBC News, audio podcast, March 30, 2010, http://news.bbc.co.uk/today/hi/today/newsid_8594000/8594561.stm.

2. Ross Gelbspan, "Beyond the Point of No Return," Grist, December 11, 2007, www.grist.org/article/beyond-the-point-of-no-return#_edn1.

3. World Bank, *World Development Indicators 2010* (Washington, DC: International Bank for Reconstruction and Development/World Bank, 2010), 91. http://data.worldbank.org/sites/default/files/wdi-final.pdf; United Nations Development Program, *Human Development Report 2006: Beyond Scarcity—Power, Poverty and the Global Water Crisis* (New York: Palgrave MacMillan, 2006), 33. http://hdr.undp.org/en/media/HDR06-complete.pdf.

4. J. S. Hacker, *The Great Risk Shift: The New Economic Insecurity and the Decline of the American Dream* (New York: Oxford University Press, 2006), 32.

5. "A Conversation with Janine Benyus," Biomimicry Guild, www.biomimicryguild.com/janineinterview.html (accessed March 8, 2011). See also Janine M. Benyus, *Biomimicry: Innovation Inspired by Nature* (New York: Harper Perennial, 1997).

6. Daniel Goleman, *Ecological Intelligence: How Knowing the Hidden Impacts of What We Buy Can Change Everything* (New York: Broadway Books, 2009).

7. "Country Scores," Environmental Performance Index 2010, Yale Center for Environmental Law and Policy, http://epi.yale.edu/Countries; David Biello, "Can Coal and Clean Air Coexist?" *Scientific American,* August 4, 2008, http://www.scientificamerican.com

/article.cfm?id=can-coal-and-clean-air-coexist-china; Elisabeth Rosenthal, "'Dirty' Energy Dwarfs Clean in China and India," *New York Times* Green (blog), February 21, 2011, http://green.blogs.nytimes.com/2011/02/21/dirty-energy-dwarfs-clean-in-china-and-india/.

8. Thomas Friedman, "Have a Nice Day," *New York Times*, September 16, 2009, www.nytimes.com/2009/09/16/opinion/16friedman.html.

9. Interview with Deborah Seligsohn by Bruce Gellerman, *Living on Earth*, National Public Radio, March 11, 2011, www.livingonearth.org/shows/shows.htm?programID=11-P13-00010.

10. *Vital Signs 2011*, 80; interview with Deborah Seligsohn on National Public Radio.

11. Al Gore, "Closing Session" (presentation at the Technology, Entertainment, and Design Conference, Monterey, California, March 2008), www.wired.com/epicenter/2008/03/al-gore-makes-i.

12. Gilbert Burnham et al., "Mortality After the 2003 Invasion of Iraq: A Cross-Sectional Cluster Sample Survey," *The Lancet* 368, no. 9545 (October 2006): 1421–1428; Eisenhower Study Group, "The Costs of War Since 2001: Iraq, Afghanistan, and Pakistan," Providence, RI: Watson Institute for International Studies, 2012.

13. Beatriz Stolowicz, "The Latin American Left: Between Governability and Change," in *The Left in the City*, ed. Daniel Chavez and Benjamin Goldfrank (London: Latin American Bureau, 2004), citing "Desarrollo más allá de la economía" (Inter-American Development Bank, Washington, DC, September 2000), 180.

14. L. Conradt and T. J. Roper, "Group Decision-Making in Animals," *Nature* 421 (January 2003): 155–157. www.cs.princeton.edu/~chazelle/courses/BIB/group-decision-making.pdf.

15. Anita Williams Woolley et al., "Evidence for a Collective Intelligence Factor in the Performance of Human Groups," *Science* 330, no. 6004 (October 2010): 686–688.

16. Richard Wilkinson and Kate Pickett, *The Spirit Level: Why More Equal Societies Almost Always Do Better* (London: Allan Lane, 2009), 73–87, 129–156.

17. Michael Kumhof and Romain Ranciere, "Inequality, Leverage and Crises, International Monetary Fund" (IMF Working Paper WP/10/268, November 2010), www.imf.org/external/pubs/ft/wp/2010/wp10268.pdf.

18. Emmanuel Saez, "Striking It Richer: The Evolution of Top Incomes in the United States (Updated with 2008 Estimates)" (working paper, Department of Economics, University of California, Berkeley, July 17, 2010), 5. www.econ.berkeley.edu/~saez/saez-UStopincomes-2008.pdf.

19. Interview with Naomi Klein by Amy Goodman, "My Fear Is That Climate Change

Is the Biggest Crisis of All," Znet.com, March 10, 2011, www.zcommunications.org/my-fear-is-that-climate-change-is-the-biggest-crisis-of-all-by-naomi-klein.

20. Chalene Helmuth, *Culture and Customs of Costa Rica* (Westport, CT: Greenwood Press, 2000), 22–25.

21. John A. Booth, *Costa Rica: Quest for Democracy* (Boulder, CO: Westview Press, 1998), 48, 71.

22. Charles D. Ameringer, *Democracy in Costa Rica* (New York: Praeger, 1982), 51.

23. Booth, *Costa Rica*, 73.

24. Luis Antonio Sobrado González and Hugo Picado León, "A Comparative Latin American Perspective of Political and Electoral Reform Experiences in Colombia, Costa Rica and Mexico" (Spanish), International Institute for Democracy and Electoral Assistance, 2010, www.idea.int/publications/comparative_la_perspective/upload/experiencias_de_reforma_2.pdf.

25. "Election Day Costa Rica's Most Important Holiday," Cupotico.com, www.cupotico.com/info/General/Costa_Rica_most_important_holiday_-_Election_Day.html; Electoral Commission, *Compulsory Voting Around the World* (London: Electoral Commission, 2009), 8, 24. www.electoralcommission.org.uk/__data/assets/electoral_commission_pdf_file/0020/16157/ECCompVotingfinal_22225-16484__E__N__S__W_.pdf.

26. Helmuth, *Culture and Customs of Costa Rica*, 12.

27. Alain de Janvry and Elisabeth Sadoulet, "Land Reforms in Latin America: Ten Lessons toward a Contemporary Agenda" (presentation, World Bank's Latin American Land Policy Workshop, Pachuca, Mexico, June 14, 2002), 12, http://are.berkeley.edu/~sadoulet/papers/Land_Reform_in_LA_10_lesson.pdf.

28. "Country Scores," Yale Center for Environmental Law and Policy; "Voter Turnout Database," International Institute for Democracy and Electoral Assistance, www.idea.int/vt/viewdata.cfm#prebuilt=yes&countries=54,109,197,42&types=parl,pres&fields=vt,vap,pop,valid&logurl=http%3A%2F%2Fwww.idea.int%2Fvt%2Fviewdata.cfm&timemode=all&quickView=false&qid=973129&d=1&h=13&m=57 (accessed January 17, 2011).

29. Abigail Hauslohner, "Out of a Village in Egypt: Portrait of a Facebook Rebel," *Time*, February 20, 2011, www.time.com/time/world/article/0,8599,2049646,00.html.

30. Jon Mooallem, "The End Is Near! (Yay!)," *New York Times Magazine*, April 19, 2009, 35.

31. *BBMG Conscious Consumer Report: Redefining Value in a New Economy* (New York: BBMG, 2009).

32. Co-Operative Bank, *Good with Money, Ten Years of Ethical Consumerism* (Manchester, UK: Co-Operative Bank, 2010), www.goodwithmoney.co.uk/assets/Ethical-Consumerism-Report-2009.pdf.

33. For more information on how to join in, visit Carrotmob at carrotmob.org.

34. Thomas Geoghegan, "Consider the Germans," *Harper's*, March 2010, www.harpers.org/archive/2010/03/0082859?redirect=2101317980.

35. "Collective Bargaining," Worker-Participation.eu, http://www.worker-participation.eu/National-Industrial-Relations/Across-Europe/Collective-Bargaining2 (accessed May 31, 2011); Organization for Economic Co-Operation and Development, *OECD Employment Outlook: Moving Beyond the Jobs Crisis* (Paris: OECD, 2010), 68, 74–75, http://browse.oecdbookshop.org/oecd/pdfs/browseit/8110081E.PDF.

36. Center for Responsive Politics, "Lobbying Database," OpenSecrets.org, www.opensecrets.org/lobby/index.php (accessed January 3, 2011).

37. This is calculated from Center for Responsive Politics, "Lobbying Spending Database," OpenSecrets.org, www.opensecrets.org/lobby/list_indus.php (accessed January 13, 2011).

38. "Toxic 100 Air Polluters Index," Political Economy Research Institute at the University of Massachusetts Amherst, www.peri.umass.edu/toxic_index (last updated March 14, 2011).

39. Greenpeace, *Koch Industries Secretly Funding the Climate Denial Machine* (Washington, DC: Greenpeace, 2010), 6. Available at www.greenpeace.org/usa/Global/usa/report/2010/3/koch-industries-secretly-fund.pdf; Jane Mayer, "Covert Operations: The Billionaire Koch Brothers Who Are Raging Against Obama," *New Yorker*, August 30, 2010, www.newyorker.com/reporting/2010/08/30/100830fa_fact_mayer#ixzz1GgSAe397.

40. "BBC World Service Poll," BBC News, 2007. http://news.bbc.co.uk/2/shared/bsp/hi/pdfs/09_11_2007bbcpollclimate.pdf.

41. "All Lease Offerings," Bureau of Ocean Energy Management, Regulation, and Enforcement, Department of the Interior, www.gomr.boemre.gov/homepg/lsesale/swiler/Table_1.pdf (accessed February 18, 2011).

42. "Editorial Memorandum: The Solution to Pay-to-Play Politics," Common Cause, news release, January 12, 2009, www.commoncause.org/site/apps/nlnet/content2.aspx?c=dkLNK1MQIwG&b=4773613&ct=6492471.

43. "Political Ad Spending Doubled in 2012, More Drastically in Battleground States," *PBS NewsHour*, September 24, 2012, www.pbs.org/newshour/bb/politics/july-dec12/campaign_09-24.html; Center for Responsive Politics, "2012 Outside

Spending, by Super PACs," OpenSecrets.org, www.opensecrets.org/outsidespending
/summ.php?cycle=2012&chrt=V&disp=O&type=A (accessed November 29, 2012).

44. Center for Responsive Politics, "Real Disclosure of Donors Funding Independent Political Advertisements," OpenSecrets.org, www.opensecrets.org/action/issues/real -disclosure-donors-adverts/ (accessed November 29, 2012).

45. "Bill Summary and Status: 111th Congress (2009-2010) S. 3628," Library of Congress: THOMAS, http://thomas.loc.gov/cgi-bin/bdquery/z?d111:SN03628 (accessed January 17, 2011).

46. "Election Week Polling on Fair Elections Act and Money in Campaigns," Lake Research Partners, news release, November 4, 2010, http://fairelectionsnow.org/sites /default/files/11-2010-fairelections-polling.pdf

47. Launched in 2012, Represent Us, www.represent.us, is a campaign of United Republic.

48. "Public Financing of Campaigns: An Overview," National Conference of State Legislatures (NCSL), www.ncsl.org/default.aspx?tabid=16591 (accessed January 6, 2010).

49. "Rep. Hinck Looks to Expand E-Waste Law," Maine House Democrats, news release, February 12, 2009, www.maine.gov/tools/whatsnew/index.php?topic=House Dems+News&id=68151&v=Article; "Testimony in Support of LD 365, 'An Act to Enhance Maine's Electronic Waste Law,'" Natural Resources Council of Maine, news release, March 10, 2009, www.nrcm.org/news_detail.asp?news=2887.

50. State Legislation," Electronics TakeBack Coalition, http://www.electronic-stakeback.com/promote-good-laws/state-legislation/

51. Calculated by Public Citizen from data in "Spending in the 2008 Election," Financing the 2008 Election, ed. David Magelby and Anthony Corrado (Brookings Institution Press, Washington, D.C., 2010), http://www.citizen.org/documents/presidential -election-public-financing-facts-numbers.pdf

52. "Public Financing of Campaigns: An Overview."

53. To learn more, visit Represent Us at www.represent.us; the Grassroots Democracy Act, http://www.johnsarbanes.com/petition/grassroots-democracy-act; for the Fair Elections Now Act, www.fairelectionsnow.org

54. Chico Whitaker, personal communication with author, March 2011.

55. L. Conradt and T. J. Roper, "Group Decision-Making in Animals," Nature 421 (January 2003): 155-157. www.cs.princeton.edu/~chazelle/courses/BIB/group-decision -making.pdf.

56. James Surowiecki, *The Wisdom of Crowds* (New York: Doubleday, 2004), 30, 38, 49, 58, 60, 65.

57. For more, see our *Doing Democracy* handbook on the Small Planet website at www.smallplanet.org/sites/default/files/doing_democracy_10_practical_arts _handbook.pdf.

58. Michel Pimbert et al., *Democratising Agricultural Research for Food Sovereignty in West Africa* (London: International Institute for Environment and Development, 2010), 4–5. http://pubs.iied.org/pdfs/14603IIED.pdf.

59. Pimbert et al., *Democratising Agricultural Research for Food Sovereignty in West Africa*, 4–9.

60. James S. Fishkin, *When the People Speak* (Oxford: Oxford University Press, 2009), 124.

61. R. L. Lehr et al., *Listening to Customers: How Deliberative Polling Helped Build 1,000 MW of New Renewable Energy Projects in Texas* (Golden, CO: National Renewable Energy Laboratory, 2003), 3, 10–11. http://cdd.stanford.edu/polls/energy/2003/ renewable_energy.pdf.

62. Stefan Theil, "Greenest Nation: A Laggard No Longer, America Could Soon Out-Innovate Europe and Japan," *Newsweek*, March 2, 2009, www.newsweek.com/2009/ 02/21/greenest-nation.html#.

63. America*Speaks*, *Our Budget, Our Economy: Finding Common Ground on our Fiscal Future* (Washington, DC: America*Speaks*, 2010), http://usabudgetdiscussion.org/ wp-content/uploads/2010/07/FindingCommonGround072710.pdf.

64. "Making Every Vote Count: The Case for Electoral Reform in British Columbia," British Columbia Citizen's Assembly on Electoral Reform, December 2004, www .citizensassembly.bc.ca/resources/final_report.pdf.

65. Dennis Pilon, "The 2005 and 2009 Referenda on Voting System Change in British Columbia," *Canadian Political Science Review* 4, nos. 2–3 (2010): 80. http://ojs .unbc.ca/index.php/cpsr/article/viewDownloadInterstitial/251/301.

66. "Folkbildning in Sweden," Swedish National Council of Adult Education, www.folkbildning.se/Folkbildning/Oversattningar/English-translations (accessed February 27, 2011).

67. "Profiles of Successful Dialogue-to-Change Programs Strengthening Neighbor-hoods," Everyday Democracy, May 14, 2008, www.everyday-democracy.org/en/Article .295.aspx; Nick Connell, Everyday Democracy, personal communication with author, June 2008.

68. "June 2010 Solid Waste Update," SedgwickCounty.org, www.sedgwickcounty.org/environment/solid_waste/Solid%20Waste%20Plan%20Update%20June%202010.pdf, 8.

69. Gianpaolo Baiocchi, "Porto Alegre: The Dynamism of the Unorganized," in *The Left in the City*, ed. Daniel Chavez and Benjamin Goldfrank (London: Latin American Bureau, 2004), 53.

70. Baiocchi, "Porto Alegre"; Tiago Peixoto, "Brazil and Argentina: From Participatory Budgeting to e-Participatory Budgeting," *Intergovernmental Solutions Newsletter: Engaging Citizens in Government* (fall 2009): 23–24.

71. Hille Hinsberg et al., "My Better Estonia," in *Intergovernmental Solutions Newsletter: Engaging Citizens in Government* (fall 2009): 20–21.

72. Hinsberg et al., "My Better Estonia"; Cristiano Faria and Tiago Peixoto, "Participatory Lawmaking in Brazil," *Intergovernmental Solutions Newsletter: Engaging Citizens in Government* (fall 2009): 22.

73. Mark R. Warren, *Faith-Based Community Organizing: The State of the Field* (Jericho, NY: Interfaith Funders, 2001), www.nhi.org/online/issues/115/Warren.html; "About PICO: History," PICO National Network, www.piconetwork.org/about?id=0006 (accessed March 3, 2011).

74. Edward W. Barrett, *Truth Is Our Weapon,* Chap. 10: "The Problem of Words" (New York: Funk & Wagnalls Co., 1953).

75. Abigail Hauslohner, "Out of a Village in Egypt: Portrait of a Facebook Rebel," Time.com, February 20, 2011, www.time.com/time/specials/packages/article/0,28804,2045328_2045333_2049646,00.html.

76. David D. Kirkpatrick and David E. Sanger, "A Tunisian-Egyptian Link That Shook Arab History," *New York Times,* February 13, 2011, www.nytimes.com/2011/02/14/world/middleeast/14egypt-tunisia-protests.html?pagewanted=1&_r=1&sq=Egypt%2018%20days%20youth&st=cse&scp=5&adxnnlx=1299866537-C8TDM04SF/GmRZMSkmzFeQ.

## An Invitation

1. Richard Kostelanetz, *Conversing with Cage* (New York: Routledge, 2003), 221.

2. Theodore Zeldin, *An Intimate History of Humanity* (New York: Random House, 1994), 197.

3. Donella Meadows, "The Forest Is More Than a Collection of Trees," *The Donella Meadows Archive* (Hartland, VT: Sustainability Institute, 2010), www.sustainer.org/dhm_archive/index.php?display_article=vn806rootgrafted.

4. Paul Stamets, *Mycelium Running: How Mushrooms Can Help Save the World* (Berkeley, CA: Ten Speed Press, 2005), 2.

5. John Naish, "We Have Enough," *Resurgence* 159 (March–April 2010): 17.

6. Donald F. Behan and Samuel H. Cox, "Obesity and Its Relation to Mortality and Morbidity Costs," Society of Actuaries, December 2010, http://66.216.104.121/files/pdf/research-2011-obesity-relation-mortality.pdf, 59; Michael J. McGinnis and William H. Foege, "The Immediate vs. the Important," *Journal of the American Medical Association* 291, no. 10 (2004): 1263–1264.

7. Jill Reedy and Susan Krebs-Smith, "Dietary Sources of Energy, Solid Fats, and Added Sugars Among Children and Adolescents in the United States," *Journal of the American Dietetic Association* 110, no. 10 (October 2010): 1477–1484.

8. Paul M. Johnson and Paul J. Kenny, "Dopamine D2 Receptors in Addiction-Like Reward Dysfunction and Compulsive Eating in Obese Rats," *Nature Neuroscience* 13 (March 2010): 635–641; Laura Sanders, "Junk Food Turns Rats into Addicts," *Science News* 176, no. 11 (November 2009): 8.

9. David Kessler, *The End of Overeating: Taking Control of the Insatiable American Appetite* (Emmaus, PA: Rodale, 2009), 21.

10. Catherine Rampell, "On School Buses, Ad Spaces for Rent," *New York Times*, April 15, 2011, www.nytimes.com/2011/04/16/business/media/16buses.html.

11. Jeffrey Levi et al., *F as in Fat: How Obesity Policies Are Failing America* (Washington, DC: Robert Wood Johnson Foundation, 2008), 20. http://healthyamericans.org/reports/obesity2008/Obesity2008Report.pdf.

12. Juliet B. Schor, *The Overworked American: The Unexpected Decline of Leisure* (New York: Basic Books, 1993), 29.

13. Karine Spiegel et al., "Brief Communication: Sleep Curtailment in Healthy Young Men Is Associated with Decreased Leptin Levels, Elevated Ghrelin Levels, and Increased Hunger and Appetite," *Annals of Internal Medicine* 141, no. 11 (2004): 846–850. www.annals.org/content/141/11/846.full.pdf+html.

14. Trenton G. Smith, Christiana Stoddard, and Michael G. Barnes, "Why the Poor Get Fat: Weight Gain and Economic Insecurity," *Forum for Health Economics and Policy* 12, no. 2 (2009): 12.

15. Dennis Olson, "Below-Cost Feed Crops: An Indirect Subsidy for Industrial Animal Factories," Institute for Agriculture and Trade Policy, June 2006, http://www.agobservatory.org/library.cfm?refid=88122; for this basic thesis, see Frances Moore Lappé, *Diet for a Small Planet* (New York: Ballantine, 1991). See also Michael Pollan, "You Are What You Grow," *New York Times Magazine*, April 17, 2007.

16. "The United States Summary Information," Farm Subsidies Data Base, http://farm.ewg.org/region?fips=00000&regname=UnitedStatesFarmSubsidySummary (accessed

March 1, 2011); Doug Henwood (editor, Left Business Observer), personal communication with author, February 24, 2011. Calculations based on Bureau of Labor Statistics data on the nominal value of the minimum wage, deflated by the consumer price index, rebased to February 2011 dollars.

17. Norman Uphoff, "Features of the System of Rice Intensification (SRI) Apart from Increases in Yield," Cornell International Institute for Food, Agriculture and Development (CIIFAD), May 2005, http://ciifad.cornell.edu/sri; Africare, Oxfam America, and World Wildlife Fund (WWF), *More Rice for People, More Water for the Planet* (Hyderabad, India: WWF–International Crops Research Institute for the Semi-Arid Tropics Project, 2010), 12. www.oxfamamerica.org/files/more-rice-for-people-more-water-for-the-planet-sri.pdf. See also "System of Rice Intensification (SRI)," CIIFAD, http://ciifad.cornell.edu/SRI/advant.html.

18. "A Conversation with Norman Uphoff, Advisor to Nourishing the Planet," Ground Report, May 14, 2010, www.groundreport.com/World/A-Conversation-with-Norman-Uphoff-Advisor-to-Nouri/2923359.

19. C. Shambu Prasad, *System of Rice Intensification in India: Innovation History and Institutional Challenges* (Andhra Pradesh, India: WWF-ICRISAT, 2006), 14. www.crispindia.org/docs/SRI%20in%20India%20innovation%20and%20institutions.pdf.

20. For more about the work of the Ecologic Development Fund, visit its website at Ecologic.org.

21. Lydia Polgreen, "Mali's Farmers Discover a Weed's Potential Power," *New York Times*, September 9, 2007, www.nytimes.com/2007/09/09/world/africa/09biofuel.html?pagewanted=print.

22. Pere Ariza-Montobbio et al., "The Political Ecology of Jatropha Plantations for Biodiesel in Tamil Nadu, India," *Journal of Peasant Studies* 37, no. 4 (October 2010): 875–897.

23. Helen Burley and Adrian Bebb, eds., *Losing the Plot: The Threats to Community Land and the Rural Poor Through the Spread of Biofuel Jatropha in India* (Brussels: Friends of the Earth Europe, 2009), 20, www.foeeurope.org/agrofuels/jatropha_in_india.pdf.

24. United Nations Environment Program–United Nations Conference on Trade and Development (UNEP-UNCTAD) Capacity-Building Task Force on Trade, Environment, and Development, *Organic Agriculture and Food Security in Africa* (New York: United Nations, 2008), 11–16. www.unctad.org/en/docs/ditcted200715_en.pdf.

25. UNEP-UNCTAD Capacity Building Task Force on Trade, Environment and Development, *Organic Agriculture and Food Security in Africa*, 13–14.

26. *Rural Women and Food Security: Current Situation and Perspectives* (Rome: FAO, 1998). www.fao.org/DOCREP/003/W8376E/W8376E00.HTM; UNEP-UNCTAD Capacity Building Task Force on Trade, Environment and Development, *Organic Agriculture and Food Security in Africa*, 13.

27. Catherine Badgley et al., "Organic Agriculture and the Global Food Supply," *Renewable Agriculture and Food Systems* 22 (2007): 86–108.

28. M. S. Swaminathan, "An Evergreen Revolution," *Crop Science* 46, no. 5 (2006): 2293–2303.

29. "The Billion Tree Campaign—Growing Green," United Nations Environment Program, www.unep.org/billiontreecampaign/index.asp (accessed March 15, 2011).

30. Abebech Tamene, "Ethiopia's Promising Environmental Campaigns," Ethiopian News Agency, www.ena.gov.et/Link4/Ethiopia_Promising_Envtal_Campaigns.pdf; interview with President Girma Wolde Giorgis by Satinder Bindra, "Transforming Ethiopia," in "Green Economy: The New Big Deal," *Our Planet*, February 2009, www.unep .org/pdf/ourplanet/2009/feb/en/OP-2009-02-en-FULLVERSION.pdf, 19.

31. To register a tree you plant, visit the Billion Tree Campaign at www.unep.org/ billiontreecampaign.

32. Energy Sector Management Assistance Program (ESMAP), "Good Practices in City Energy Efficiency: Bogota, Columbia—Bus Rapid Transit for Urban Transport" (ESMAP, Energy Efficient Cities Initiative, November 2009), 1, 9. www.esmap.org/esmap/sites/ esmap.org/files/CS_Bogota_020310_0.pdf.

33. Elizabeth Rosenthal, "Buses May Aid Climate Battle in Poor Cities," *New York Times*, July 10, 2009, www.nytimes.com/2009/07/10/world/americas/10degrees.html.

34. Maria Victoria Llorente and Angela Rivas, *Case Study Reduction of Crime in Bogotá: A Decade of Citizen's Security Policies* (Washington, DC: World Bank, 2005), 5. Available at http://siteresources.worldbank.org/EXTLACREGTOPURBDEV/Resources/ 841042-1219076931513/5301922-1250717140763/Bogota.pdf.

35. Alan Gilbert, "Bus Rapid Transit: Is Transmilenio a Miracle Cure?" *Transport Reviews* 28, no. 4 (2008): 439–467; "Bus Rapid Transit: Lessons from Latin America," Global Mass Transit Report, policy review, November 1, 2009, www.globalmasstransit .net/archive.php?id=426.

36. Chattanooga Green Committee, *The Chattanooga Climate Action Plan* (Chattanooga, TN: Chattanooga Green Committee, 2009), 15, 20, 113. www.chattanooga .gov/Final_CAP_adopted.pdf.

37. Rachel Cleetus, Steven Clemmer, and David Friedman, *Climate 2030: A National Blueprint for a Clean Energy Economy* (Cambridge, MA: Union of Concerned Scientists,

2009), 18. www.ucsusa.org/global_warming/solutions/big_picture_solutions/climate-2030-blueprint.html.

38. Federal Ministry of Economics and Technology, *Energy Concept for an Environmentally Sound, Reliable and Affordable Energy Supply* (Berlin: Federal Ministry of Economics and Technology, 2010), 5. www.bmwi.de/English/Redaktion/Pdf/energy-concept,property=pdf,bereich=bmwi,sprache=en,rwb=true.pdf.

39. "Ethical Funds Can Make Good Investment Returns," UK Investment Advice, February 2, 2007, www.ukinvestmentadvice.co.uk/ethical-funds.htm. For the banana estimate, see Harriet Lamb, "Power to Producers," *Resurgence* 259 (March–April 2010): 35.

40. For background, visit Friends of the MST at www.mstbrazil.org.

41. Markus Laegel, "Berlin Wall Fell As Young People Prayed," 24-7 Prayer, March 25, 2005, www.24-7prayer.com/articles/661.

42. Laegel, "Berlin Wall Fell As Young People Prayed."

43. Solomon E. Asch, "Opinions and Social Pressure," *Scientific American* 193, no. 5 (1955): 31–35.

44. James Surowiecki, *The Wisdom of Crowds* (New York: Doubleday, 2004), 65.

45. Simone Schnall et al., "Social Support and the Perception of Geographical Slant," *Journal of Experimental Social Psychology* 44 (2008): 1253.

46. On the discovery of mirror neurons, see G. Di Pellegrino et al., "Understanding Motor Events: A Neurophysiological Study," *Experimental Brain Research* 91 (1992): 176–180. See also Sandra Blakeslee, "Cells That Read Minds," *New York Times*, January 10, 2006, www.nytimes.com/2006/01/10/science/10mirr.html.

47. See the Warren Company at www.warrenco.com.

48. Jared Diamond, *Collapse: How Societies Choose to Fail or Succeed* (New York: Viking Penguin, 2005), 79–119.

49. Erich Fromm, *The Heart of Man* (New York: Harper & Row, 1964), 26–27.

50. Susan Osborn (singer-songwriter, www.susanosborn.com), e-mail communication with the author, December 2010.

# INDEX